BATRACOLOGY

BATRACOLOGY

Dr. Digumarti Bhaskara Rao
M.Sc., M.A., M.A., M.Ed., Ph.D.
R.V.R. College of Education
Guntur (A.P.)

1995
DISCOVERY PUBLISHING HOUSE
NEW DELHI—110 002

First Published—1995

ISBN 81-7141-279-3

Published by :

Discovery Publishing House
4594/9, Darya Ganj,
New Delhi—110 002 (INDIA)
Phone : 3279245

Laser Typesetting by :
Allied Computers,
Karnal

Printed at : Tarun Offset Printers
518/14, Anup Market, Maujpur, Delhi-53

Preface

The frogs are very commonly found in our viscinity. We hear the croaking or singing of male frogs to attract the females for mating during their breeding season. Some times we enjoy it and at other times we bear it.

The frogs have great scientific interest. Students dissect the frogs in laboratories to study the external and internal structures for their examination purposes. On the other hand, many scientists conduct experiments with frogs in sophisticated laboratories to study the behaviour, to measure the effects of drugs, to know the anatomy, to learn the physiological activities, to understand the hereditary principles, so on and so forth.

The frogs are studied as a part of general Zoology. Though many students of +2, under-graduate, post-graduate and research on one side and the anatomists, physiologists, biochemists, biophysicists, biotechnologists, morphologists, myologists, nephrologists, neurologists, osteologists, bionomicists, cytogeneticists, ethologists, etc., on the other side study frogs in every nook and corner of the world, there is no single separate book around the globe explaining the various phenomena of frog. To meet this requirement, a small effort has been put in and the result is this tiny book—BATRACOLOGY—on frogs.

The frog is explained in this book under various heads with illustrations to give a brief, but authentic, information to the interested clientele. The knowledge about the frog is a generalized one, as there are certain exceptions due to 2,000 species of frogs are known to mankind. The author hopes that this book will be of great use to the students and scientists. Constructive criticism will be highly appreciated.

Dr. D. Bhaskara Rao

Acknowledgements

I express my thanks to

Dr. L. Rathaiah

Mr. P. Koteswara Rao

Mr. M. Nageswara Rao

Mr. G. Nageswara Rao

Dr. K.R.S. Sambasiva Rao

Mr. N.S.R. Prasad

Mr. Y. Jaya Sankar Prasad

Mr. V. Venkateswara Rao

Mr. P. Venkata Rao

Mr. D. Venkata Rao

Ms. D. Vijaya Lakshmi

Mr. M. Radha Krishna Murthy

Mr. T.R. Wasan

for their academic support.

D. B. RAO

Contents

1

Introduction

We hear him but we can't see him. His peculiar singing is most attractive on a damp night, when the air is with lots of moisture. He sits still and sings hours together without any interruption. We cannot see him as he can change the colour of his body to match it to the surroundings. He is a frog, a small animal. Some of the frogs stay near the ponds, some of them live on trees and some others live in fresh waters, but majority of them lead an amphibiaus life. The study of frogs is called BATRACOLOGY.

THE POSITION

The frogs belong to the class Amphibia of the Animal Kingdom. The amphibians may be easily identified by their moist, glandular skin and this skin contrasts with the scaly skin of fishes and reptiles, the animals most closely related to amphibians.

Amphibians differ from fishes in certian features. Amphibians have legs instead of fins, but not all amphibians have legs. Amphibians breathe by means of lungs instead of gills, but certain fishes also have lungs. All amphibians at some stage of their development have gills, though in most species they are lost when adulthood is reached. The amphibians known as caecilians have scales, certain fishes also lack scales.

The scaly dry skin of most reptiles distinguishes them from amphibians. Amphibians possess functional gills at some time or other, but reptiles never do, though gills exist as a transitory

A familiar Cuban tree frog. It has expanded adhesive toe pads typical of many tree frogs. These pads help the frogs be agile tree climbers.

The common edible frog is usually found on the banks of streams in central Europe and Asia. Its hindlegs are eaten as delicacies.

The leopard frog, Rana pipiens*, is common throughout North America. These frogs are widely used in biological research.*

The bullfrog, Rana catesbeiana*, is a large frog native to North America. Its deep-voiced croaks are frequently heard during the summer.*

stage in the development of the reptile embryo. The amphibians always lay eggs in water or very moist places or on trees, but the reptiles lay eggs on land only. Amphibians never develop an amnion, an embryonic membrane, found in reptiles.

THE EVOLUTION

Crossopterygians, the primitive fishes, were the fore runners of the amphibians. The developed fins that came to resemble primitive legs. The crossopterygians hand lungs and in many aspects they were similar to the primitive amphibians.

The South American bullfrog, about eight inches (20cm) long, does not lay its eggs in water but builds a nest with slime and leaves and lays its eggs there, surrounded by foam. The metamorphosis of tadpoles into frogs takes place in the foamy nest.

The first amphibians made their appearance some 30,00,00,000 years ago. The first known amphibian, *Ichthyostega*, possessed limbs much like those of modern amphibians. Possibly it had a some what fish-like tail and bore other resemblances to fishes.

Schreigel's green frog lives in flat country all over Japan. Its eggs look a bit like bubbles.

In the carboniferous and permian periods, 22,50,00,000 to 35,00,00,000 years ago, there were two main groups of amphibians—Lepospondyles and Labyrinthodonts. It is believed that today's salamanders and caecilians are the descendents of the Lepospondyles, while Frogs are more closely related to the Labyrinthodonts. The reptiles also evolved, almost certainly, from the Labyrinthodonts.

A South American horned frog. These large frogs lie in hiding for their prey, pouncing out to seize their victims. They eat insects, frogs (even their own kind), lizards, snakes, and small birds and mammals.

AMPHIBIOUS LIFE

We can only guess why and how the earliest amphibians moved from the fresh water habitat of their ancestors to land. The evolution of lungs would enable certain early fishes to live in warm stagnant pools, which would be defficient in oxygen. The fishes would be able to supplement the supply of oxygen by breathing it from the outer air though their lungs. When droughts dried up the pools, primitive limbs would enable the animals to travel over land to seek other water sources.

The common frog of Europe. When attacked, it sometimes puffs up its body, brings its hind legs up against its sides and puts its hands over its eyes. This may make it difficult for its attacker to grasp it.

The goliath frog of equatorial Africa is a giant among the frogs and toads. It grows to a length of 12 inches and may be as heavy as a house cat. Behind each eye it has a gland that makes a poisonous substance. This frog lives in the Cameroons and Gabon.

The first amphibians that ventured onto land also found no vertebrate rivals. It was a fresh and quite environment, far more favourable for their development than the earlier one.

The amphibians evolved into a great variety of forms once. They became partly adapted to terrestrial life. Some were grotesque creatures with broad bony heads and others resembled modern crocadiles in size, appearance and habits.

Goeldi's frog, about 1.5 inches long, lives in Peru and Ecuador. The female carries the eggs in a pouch on her back. Eventually she goes into the water and the eggs hatch into tadpoles.

The Frog

The name *frog* includes frogs and toads, which make up the **Order Anura**. These are distinguished from other amphibians by the lack of a tail in the adult. The frogs are the spriest of the amphibians.

The giant tree frog is about four inches long and lives in the jungles of South America.

Phylum	*Chordata*
Sub-phylum	*Vertebrata*
Super-class	*Gnathostomata*
Class	*Amphibia*
Order	*Anura*
Family	*Ranidae*

Frogs hop or jump from one place to other. If we were able to leap like a frog, we would probably be able to long jump more than 30 metres. Frogs have been known to jump as much as 36 times their own length and of course in the water they are excellent swimmers.

The gray tree frog, Hyla versicolor, leaps from his perch to a new branch. Its long hind legs are well adapted for hopping and leaping.

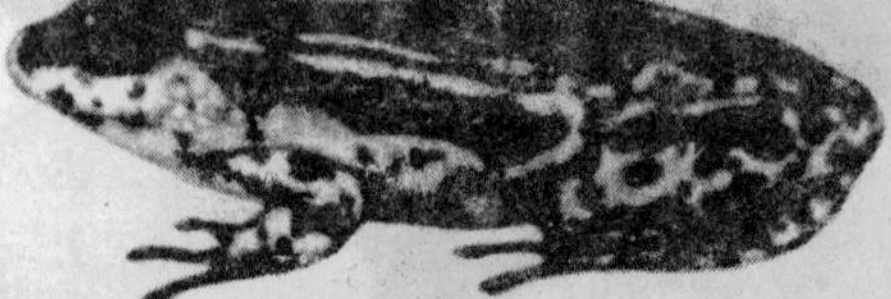

The African silver-speckled frog, about one inch long, is a leaf frog, living in African forests in the Cameroons and Congo.

Frogs are cold blooded or poikilothermic anamniotes. They possess two pairs of pentadactyle limbs which are supported by girdles and are adapted for locomotion both on land and in water.

The rare hairy frog of western Africa is about four inches long. The male develops hair-like projections on its thighs and sides during the mating season. These projections are actually long, thin outgrowths of the skin whose purpose is to help the frog breathe more oxygen as it becomes active.

Frogs are the most numerous of the three amphibian groups—frogs, salamanders and caecilians. About 2,000 species of frogs are known to date. The order name, **Anura**, means **without a tail**, and refers to a distinguishing character of these amphibian groups.

Frogs are found in most parts of the world, except where it is cold. They are almost entirely absent from Arctic and Antarctic regions and from the islands of the sea. However, a very few may be found north of the Arctic Circle and some have been introduced on oceanic islands by people. They prefer warm climates and the vast majority of species, about 80% of those known, live in the tropics.

Frogs feed on small animals like insects, worms, spiders, slugs and snails. They find their prey by sight. Insects, like flies and ants, are their favourite food. They catch them by flicking out the tongue, which is covered with sticky saliva. The movement

The rare New Zealand tailed frog, about two inches long, lives on mountain ridges far from water. It lays eggs in moist places under rocks or logs. The tadpole stage is bypassed entirely, and the eggs hatch into tiny frogs with tails. As the frogs grow they gradually lose their tails.

White's tree frog, is the most widespread of all the Australian tree frogs. It is an expert tree climber and often perches among the leaves. It feeds mainly on small insects, but at times it might swallow an entire snail.

is performed with lightning speed. The insects get trapped in the saliva and are swallowed. Frogs have got teeth, but these are not used to chew or bite their food. The teeth only serve to prevent the prey from escaping. Some frogs eat larger animals.

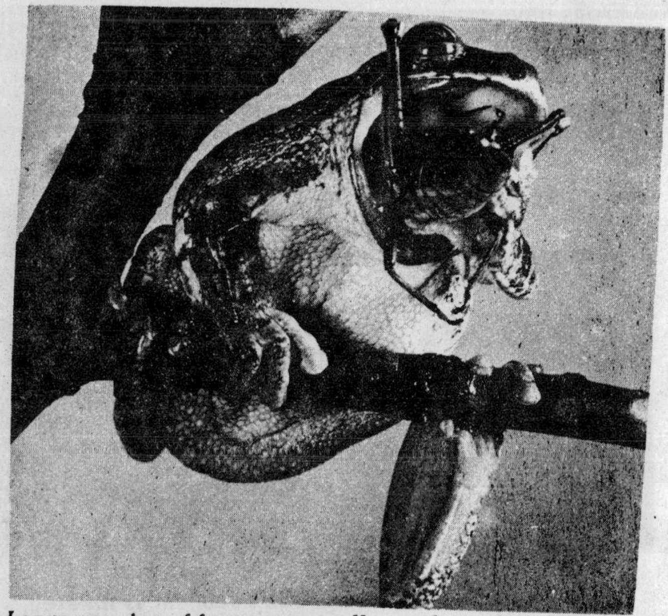

Large species of frogs can swallow relatively large prey. These animals use a long sticky tongue to catch food and get it to their mouths in a fraction of a second.

In destroying vast numbers of harmful insects, the toads inhabiting gardens and farms, do a service to mankind. One species of toad, the giant marine toad of tropical America, has been introduced in many of the islands of the Pacific in order to help control insect pests.

The common toad is found from Europe to central Asia. This nocturnal hunter has a diet of insects, snails and worms. It has rough, moist skin.

Frogs have many enemies. Birds, snakes, rats and fish are a few examples of animals that take them for food. But frogs have

An American toad singing. The vocal sac is inflated, amplifying the sounds from the vocal cords.

Wallace's flying frog is a tree frog of Southeast Asian rain forests. Its webbed feet act like parachutes when the frog jumps from branches. It does not actually fly but glides short distances.

several defence mechanisms. Most frogs are brown, grey or green or a combination of all these three colours. This, comonflage, means that the colours of their bodies merge with the colours of the surroundings. And they are very difficult to see if they stay still. Some of them can change the colour of their skin so that they look like the background.

Frog burries deep into the mud of the pond to protect itself from the severe cold in winter and scorching heat in summer. It becomes very inactive and enters into a state of dormancy in summer and winter. It carries all the vital acitivities at a slow rate to conserve energy during this state of dormancy. The inactive period or the state of dormancy during winter season is known as **hibernation**. The inactive period or the state of dormancy in summer is called **aestivation**. Glycogen is the source of energy during the dormant period. Frog exclusively performs cutaneous respiration during hibernation and aestivation. Frog excretes uric acid instead of urea during the winter sleep and summer sleep.

The clawed toad lives in ponds and lakes in central and southern Africa. When the ponds dry up it burrows into the mud until rain fills the ponds. It has claws on three of its rear toes that it uses to scratch for food.

The Asian horned toad is native to Southeast Asia. It looks like a dry, dead leaf.

Color change is accomplished by the contraction and expansion of pigment within skin cells known as melanophores. These cells are under the control of the adrenal and pituitary galnds. One would think that a frog would instinctively change its colour to match the colour of the rocks or leaves upon which it sits. This does not appear to be the case. Colour changes seem to be due to such factors as the animal's temperament and its temperature. It is not unusual to see a bright green frog sitting on a drab, gray rock.

Some frogs have poison glands in their skin. The fluids that are released from these glands are bad-smelling, bad-tasting and often very poisonous. The fumes from these poisonous glands attract the eyes, nose and skin of an attacker. This gives the frog a chance escape unharmed. The poison of Kokoi tree frogs is very strong and causes death, a few thousandths of a gram can kill a man or an animal. A poisonous frog is usually brightly coloured.

Arrow poison frogs, about 1.5 inches long, are found in Central and South America. Their brilliant colours warn other wildlife that the frogs are poisonous. Natives tip their arrows with the poison, which is secreted through the frog's skin. The male frog carries the eggs on his back. Even after the eggs hatch, the tadpoles cling to the father's back until they develop further.

This gives a warning to other animals not to eat it because it is poisonous. The Kokoi frogs, which are deadly, for instance, are brilliant red or yellow with black stripes.

Frogs and toads look very much alike and it is often difficult to identify them, but there are certain distinctive differences.

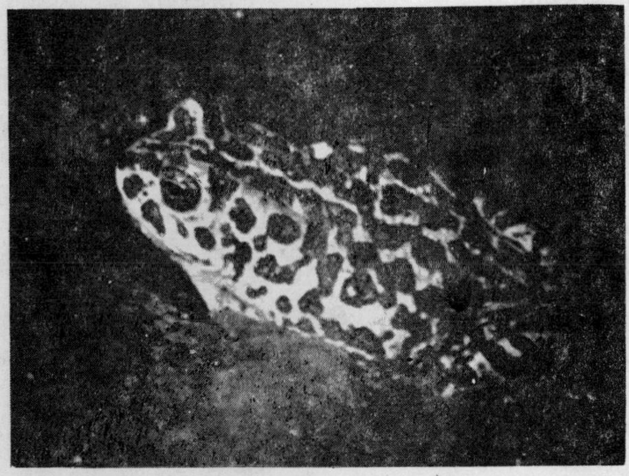

The European green toad. Toads, unlike frogs, can spend a lot of time on land. Their thick skins stop them drying up.

Differences between Frogs and Toads.

	Frogs	*Toads*
1.	Mostly aquatic.	Terrestrial.
2.	Diurnal.	Nocturnal
3.	Skin is smooth, moist and sticky.	Skin is dry, rough and warty.
4.	Maxillary teeth present on upper jaw.	Teeth are totally absent.
5.	Tongue is bifid at its free end.	Generally the tongue is not bifid.
6.	Vocal sacs are present in males.	Vocal sacs are absent in males.
7.	Vomerine teeth are present.	Vomerine teeth are absent.
8.	Lower jaw is triangular	Lower jaw is semi-circular.

Frogs	Toads
9. Bidder's organ is absent but Bidders canal is present.	Bidder's organ is present but Bidders canal is absent.
10. Parotid glands are absent.	Parotid glands are present
11. Eggs are laid in clusters.	Eggs are laid in strings.
12. Omosternum is present.	Omosternum is absent.

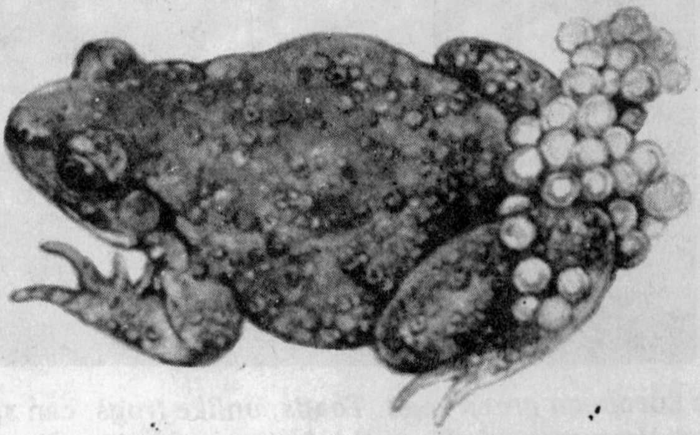

The midwife toad is noted for the male's breeding behaviour. Eggs are laid in strands, and after they are fertilized, the male wraps them around its legs and keeps them moist. In about three weeks it enters the water, where the young bite through the egg coverings and emerge to develop later into adults.

Most frog species produce immense numbers of eggs. Since comparatively few of the larvae that are hatched reach maturity, the mortality among them must be great.

Frogs that reach adulthood probably do not live more than three or four years on the average. Animals kept in captivity, however, have lived far longer than that. An American toad attained the venerable age of 31 years; a bullfrog also reached the age of 31.

The Japanese foam-nesting tree frog is green in colour. It builds its foam nest on branches overhanging the water. As the female lays the eggs, the male uses his hind legs to beat up the egg jelly into a foam that will harden to form a protective crust around the eggs. When the eggs hatch, the foam softens, and the tadpoles fall into the water.

2

External Characters

The body of frog consists of a head and trunk, but, neck and tail are absent: Absence of neck is an adaptation for swimming in water.

HEAD

The anterior most part of the body is head. It is **triangular** in shape with a slighly pointed **snout**. The tipof the snout bears a pair of ***external nostrils*** which open into the bucco-pharyngeal cavity through ***internal nostrils.***

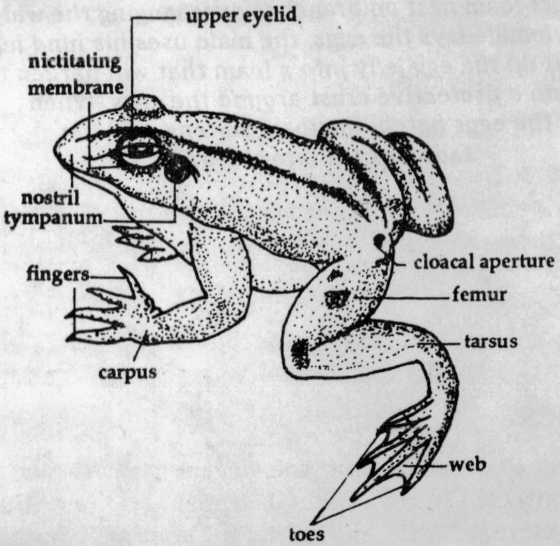

External Characters

MOUTH

A large wide mouth is situated at the anterior end of the head. Mouth is bounded by both the jaws.

EYES

A pair of large prominent eyes is found behind the external nostrils, one on each side of the head. Each eye is covered by an immovable upper eye lid and slightly movable lower eye lid. A thin transparent fold called ***nictitating membrane (third eye lid)*** is produced by the lower eye lid from its inner margin. It moves back and forth over the eye ball. Nicitating membrane keeps the eye ball moise when the frog is on land and also offers protection in water. Eye brows and eye lashes are absent in frog.

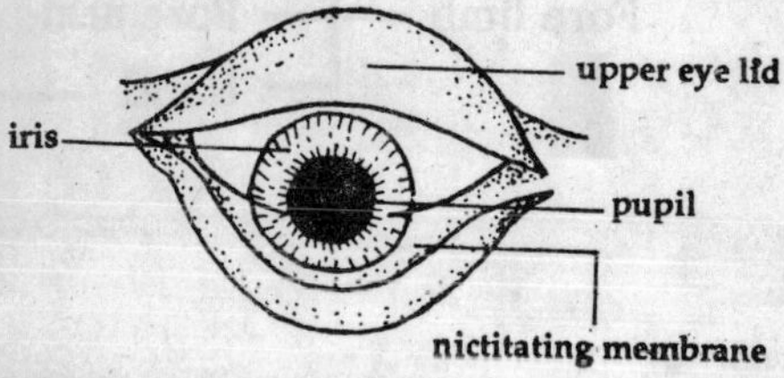

Eye

TYMPANUM

The external ear or pinna is absent in frog. There is a circular patch of skin called ***tympanic membrane (ear drum)*** behind each eye. It is tightly stretched over the cavity of middle ear. A pair of such ear drums receives the sound vibrations.

BROWSPOT

A median ***brow spot*** is present as a light coloured patch on the dorsal side of the head in the middle line between the anterior margins of the two eyes. It represents a vestigeal pineal eye.

TRUNK

The broad flattened part behind the head is trunk. The trunk bears a pair of short ***fore limbs*** and a pair of long ***hind limbs***.

FORE LIMBS

The fore limbs are shorter and weaker than the hind limbs. Each fore limb consists of 3 parts, 1. Upper arm (***brachium***), 2. forearm (***antebrachium***) and 3. hand (***manus***). The hand is further divisible into wrist (***carpus***), palm (***metacarpus***) and fingers (***digits***). There are only four fingers in each fore limb. Thumb is absent in the fore limb of frog, but a reduced ***pollex*** is present beneath the skin.

- Fore limb
  - Upper arm
  - Fore arm
  - Hand
    - Wrist
    - Palm
    - Fingers

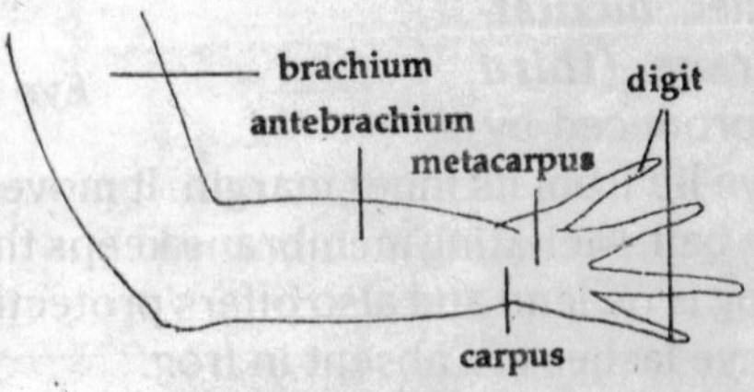

Fore limb

HIND LIMBS

The hind limbs are longer and stronger than the fore limbs. Each hind limb consists of thigh (***femur***), shank (***crus***) and foot (***pes***). The foot is divisible into ankle (***tarsus***), instep (***metatarsus***) and toes (***digits***).

- Hind limb
  - Thigh
  - Shank
  - Foot
    - Ankle
    - Instep
    - Toes

In each hind limb there are five toes. The hind limbs are adapted both for swimming and leaping. The five toes in each hind limb are connected by a thin membrane of skin known as *web* making the foot an efficient organ for swimming. Muscular cushions known as ***sub-articular pads*** are present underneath the joints of toes. The hind limbs are folded in the form of letter 'Z' in the resting position.

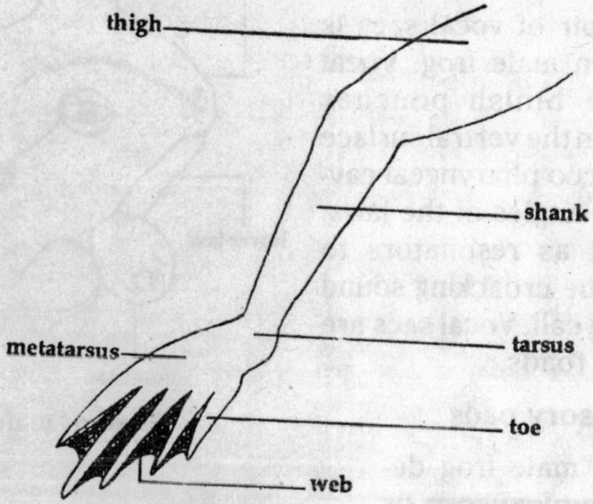

Hind limb

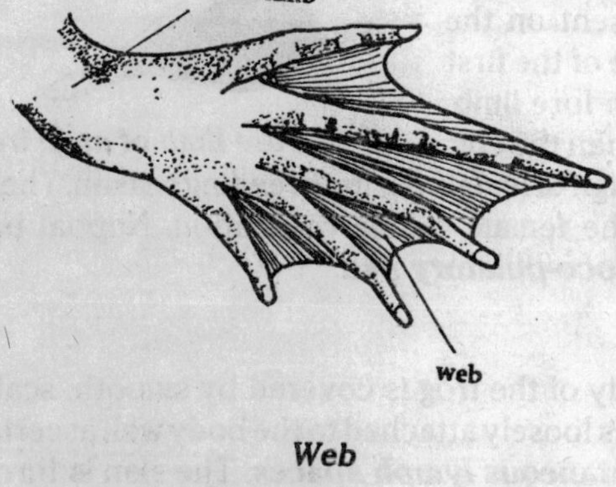

Web

CLOACAL APERTURE

It is a transverse slit located at the posterior end of the trunk. Through this opening the undigested matter, urine and sex cells pass out.

SEXUAL DIMORPHISM

Frogs are unisexual and exhibit sexual dimorphism. In male Frogs, ***vocal sacs*** and ***nuptial pads*** or ***amplexusory pads*** are present but in females they are absent.

Vocal Sacs

A pair of vocal sacs is present in male frog. Vocal sacs are bluish pouches present on the vertral surface of the bucco pharyngeal cavity at the angles of the jaws. They act as resonators to amplify the croacking sound or mating call. Vocal sacs are absent in toads.

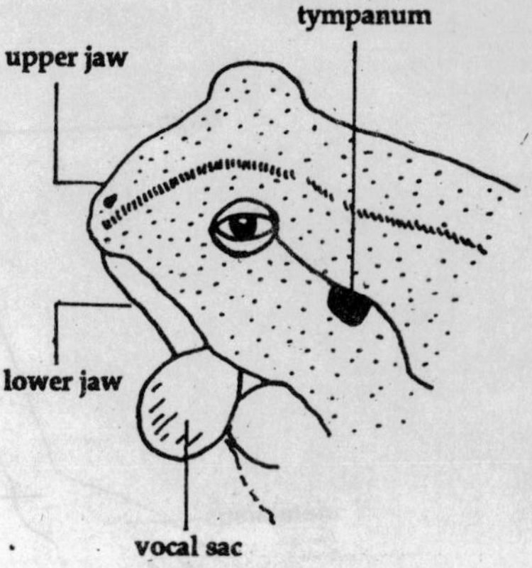

Head of male frog

Amplexusory pads

The male frog develops ***amplexusory or nuptial pads*** during the breeding season. This nuptial or amplexusory pad is a cushion-like swelling present on the lower surface of the first finger of each fore limb. It is thicker than the corresponding digit of female during breeding season. The pads help in clasping the female during copulation. Nuptial pad is also called ***pseudoco-pulatory pad***.

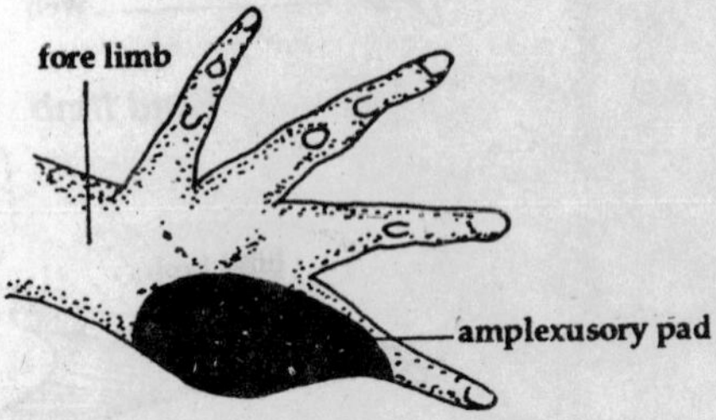

Fore limb of male frog

SKIN

The body of the frog is covered by smooth, scaleless and moist skin. It is loosely attached to the body wall at certain places due to ***subcutaneous lymph spaces***. The skin is thrown into a number of folds which extend from behind the eyes. These folded thickenings of the skin are known as ***dorsolateral dermal plĩcae.***

The colour of the skin is greenish yellow with dark spots on dorsal surface. But on ventral surface the skin is pale yellow in colour. However, frog can change the colour of its skin to suit the colour of its surroundings. This phenomenon is called ***protective colouration***. ***Intermidin or Melanocyte stimulating hormone*** brings the colour changes of the skin in adult frogs. The skin possesses mucous glands and poison glands besides chromatophores.

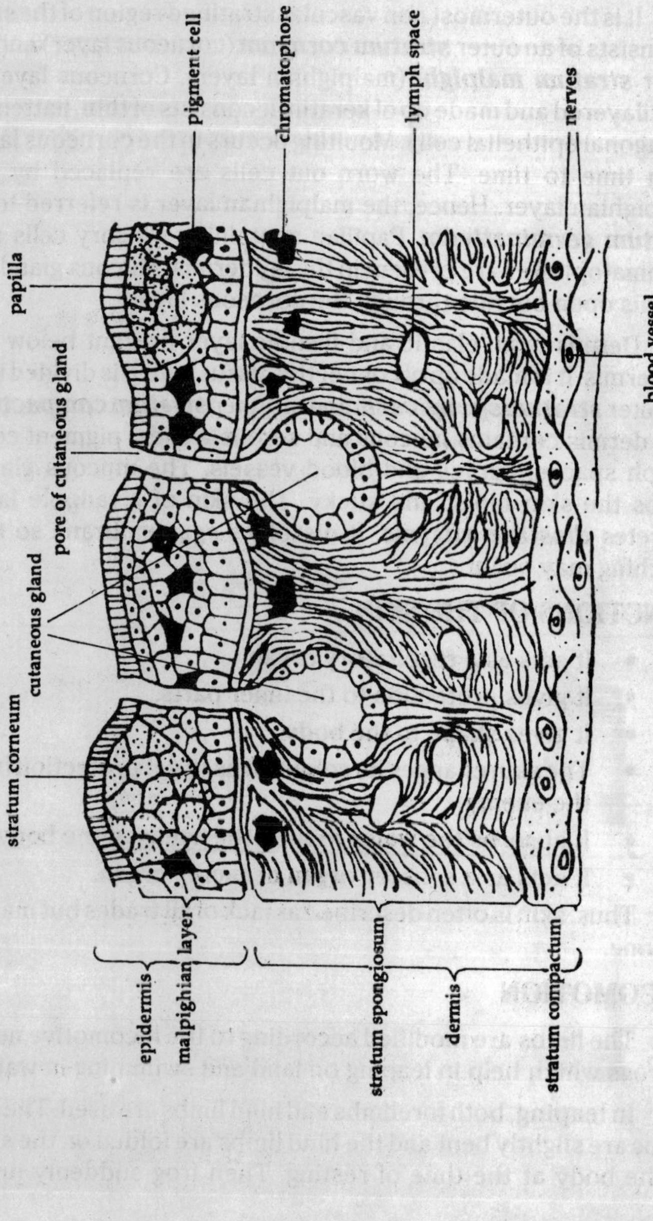
stratum corneum
cutaneous gland
pore of cutaneous gland
papilla
pigment cell
chromatophore
lymph space
nerves
blood vessel
epidermis
malpighian layer
stratum spongiosum
dermis
stratum compactum

T.S. of Skin

The skin is composed of two layers namely, ***epidermis and dermis.***

It is the outermost non vascular stratified region of the skin. It consists of an outer ***stratum corneum*** (corneous layer) and an inner ***stratum malpighi*** (malpighian layer). Corneous layer is multilayered and made up of keratin. It consists of thin, flattened, hexagonal epithelial cells. Moulting occurs in the corneous layer from time to time. The worn out cells are replaced by the malpighian layer. Hence, the malpighian layer is referred to as ***stratum germinativum.*** Papillae containing sensory cells and chromatophores are present in the epidermis. Mucous glands of dermis open outside through the corneous layer.

Dermis is an elastic and flexible layer present below the epidermis. It is made up of connective tissue which is divided into an outer ***stratum spongiosum*** and an inner ***stratum compactum.*** The dermis contains mucous and serous glands, pigment cells, lymph spaces, nerves and blood vessels. The mucous glands keeps the skin moist and sticky. The skin of a tadpole larva secretes ***diastase enzyme.*** It dissolves eggmembrane so that hatching may occur.

FUNCTIONS OF THE SKIN

- It acts as a respiratory organ.
- It gives protection to the inner parts.
- It gives shape to the body.
- The skin changes its colour so as to get protection from the enemy.
- It absorbs the water to fulfil the needs of the body.
- It acts as a sensory organ or tactile organ.

Thus, skin is often described as jack of all trades but master of none.

LOCOMOTION

The limbs are modified according to the locomotive needs of frogs which help in leaping on land and swimming in water.

In leaping, both forelimbs and hind limbs are used. The fore limbs are slightly bent and the hind limbs are folded on the sides of the body at the time of resting. Then frog suddenly jumps towards by stretching the hind limbs from the resting position. The fore limbs help by absorbing the shock while the frog is landing. The frog can cover about a distance of one meter in a single leap. ***The elongated tarsus is an adaption for leaping.***

In swimming, frog moves by alternative folding and unfolding of 'Z' shaped hind limbs. The webbed toes of the hind limbs push the water backwards and make the animal move forward.

In floating, the frog extends all of its legs and keeps only eyes and nostrils above the water surface.

In walking, limbs are lifted in a definite sequence. This is a slower process.

3

Viscera

When a frog is cut open along its mid-ventral line, organs present in the coelom (body cavity) are exposed. These different organs present in the coelom of an organism are collectively known as ***viscera***

The organs of viscera are bathed in coelomic fluid and this fluid reduces friction and avoids injury when the organs slide over one another. The different organs of the viscera are held in position by the folds of ***peritoneum***.

The viscera constitutes different organ systems which are as follows :

HEART

Heart is present midventrally in the anterior part of the body cavity. It is enveloped by a pericardium.

LUNGS

Lungs are found as elastic sacs on either side of the heart and they are enclosed by ***pleuroperitoneum.***

STOMACH

Stomach lies behind the liver on the right side of anterior part of coelom. It is a thick walled muscular organ which stores food. It is suspended from the dorsal body wall by a ***mesentery.***

SMALL INTESTINE

Small intestine is a long coiled tubular part. The anterior short 'U' shaped part of small intestine is called duodenum and the posterior elongated coiled part is known as ileum. The coils are held together by mesenteries.

LARGE INTESTINE

Large intestine or rectum is the posterior part of alimentary canal which opens into the cloaca. It is present in the posterior part of coelom.

LIVER

Liver is a bilobed dark reddish brown body lying just behind the heart. It encloses a greenish sperical gall bladder which stores the bile.

PANCREAS

Pancreas is a light yellowish thin gland lying between stomach and the duodenal loop.

SPLEEN

Spleen is a dark reddish spherical body. It is attached to the anterior part of rectum by means of a mesentery.

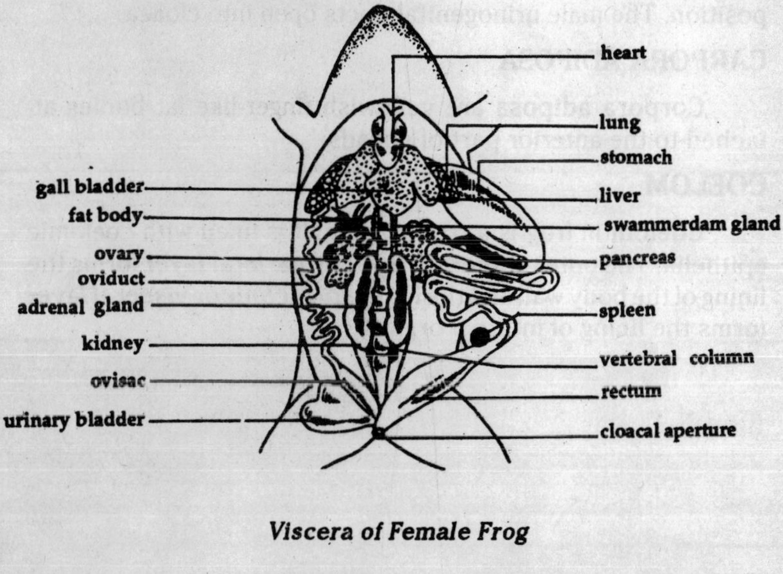

Viscera of Female Frog

KIDNEYS

Kidneys are dark reddish flattened bodies lying on either side of the dorsal aorta. The kidneys are covered by the peritoneum only on the ventral side. The somatic layer is dorsally separated from the body wall leaving a space called ***sub-vertebral lymph sinus or cisterna magna*** which encloses the kidneys. The orange coloured glands present along the midventral surface of the kidneys are ***adrenal or supra renal glands***. The ducts arising from the outer margins of kidneys are called ***ureters*** in female or ***urinogenital ducts*** in male.

URINARY BLADDER

Urinary bladder is a thin walled bilobed sac formed from the ventral surface of cloaca. It stores urine and opens into cloaca.

GONADS

In female frog, ovaries are attached to the anteroventral margins of the kidneys by ***mesovaria***. The coiled oviducts are found in the lateral part of the coelom. They run backwards from the oesophagus along the dorsal body wall and enlarge as ovisacs before opening into the cloaca.

In male, testes are present instead of ovaries in the same position. The male urinogenital ducts open into cloaca.

CARPORA ADIPOSA

Corpora adiposa are yellowish finger-like fat bodies attached to the anterior part of gonads.

COELOM

Coelom in frog is a true coelom. It is lined with coelomic epithelia. The outer ***somatick layer*** or ***parietal layer*** forms the lining of the body wall and the inner ***splanchnic*** or ***visceral layer*** forms the lining of internal organs.

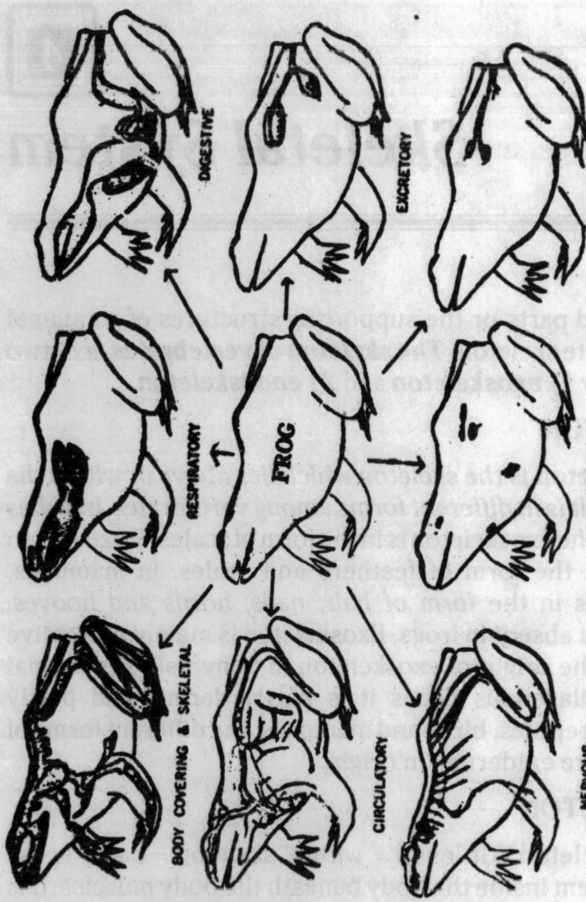

Systems of Frog

4

Skeletal System

The hard parts or the supporting structures of an animal body constitute skeleton. ***The skeleton of vertebrates is of two types namely*** 1) ***exoskeleton*** and 2) ***endoskeleton.***

EXOSKELETON

Exoskeleton *is the skeleton which lies above or within the limits of skin. It is in different forms among vertebrates.* In *fishes* and *reptires*, the exoskeleton is in the form of scales. Exoskeleton in birds is in the form of feathers and scales. In mammals, exoskeleton is in the *form of hair, nails, horns and hooves.* Exoskeleton is absent in frogs. Exoskeleton is mainly protective in function. The origin of exoskeleton in bony fishes is dermal while in cartilagenous fishes it is partly dermal and partly epidermal. In reptiles, birds and mammals, the different forms of exoskeleton are epidermal in origin.

ENDOSKELETON

Endoskeleton (Gr. endo = within, skeleton = hard) is the skeleton present inside the body beneath the body muscles. It is mesodermal in origin. Endoskeleton is made of cartilage and bone. Initially it is laid down in the form of cartilage in the embryonic stage. Some of it remains as cartilage throughout life. But most of the cartilage is replaced by bones. The bones are of two types based on the nature of development, 1. Cartilage bones 2. Dermal bones.

Cartilage or replacing bones are formed by the ossification of pre-existing cartilage of embryonic stage. Most of the bones of appendicular skeleton are replacing bones.

Dermal or membrane bones are formed by the ossification of connective tissue of dermis. Such bones invest the cartilages and are also known as ***investing bones***. These bones does not pass through cartilage stage. Bones of the axial skeleton are mostly dermal in nature.

In frog there is no exoskeleton. Endoskeleton alone is present. It is made up of both cartilage and dermal bones.

Endoskeleton forms an internal framework which gives definite shape and support to the body. It protects the delicate internal organs like brain, heart, lungs and sense organs. It offers surface for the attachment of muscles. Endoskeleton forms levers with muscles which facilitate the movement of parts of the body or the entire body.

Endoskeleton is divided into two parts,

1. Axial and 2. Appendicular skeletons.

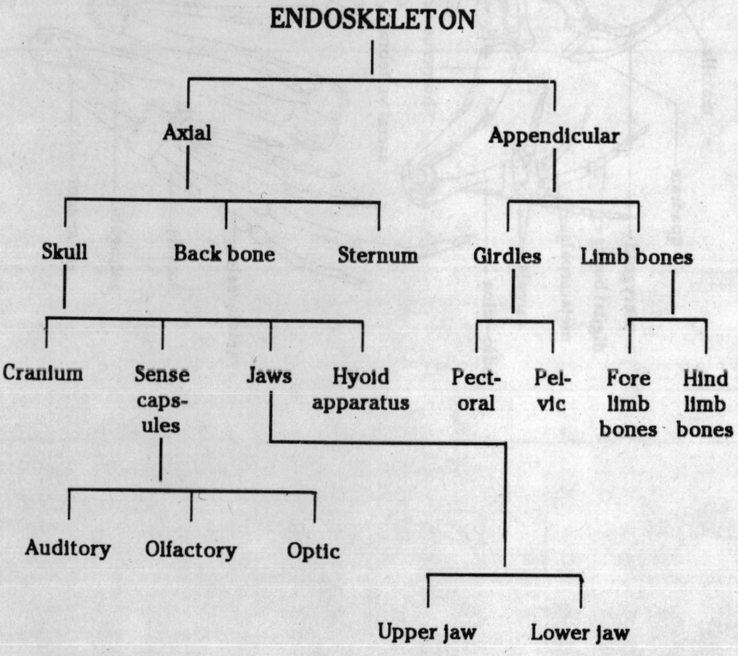

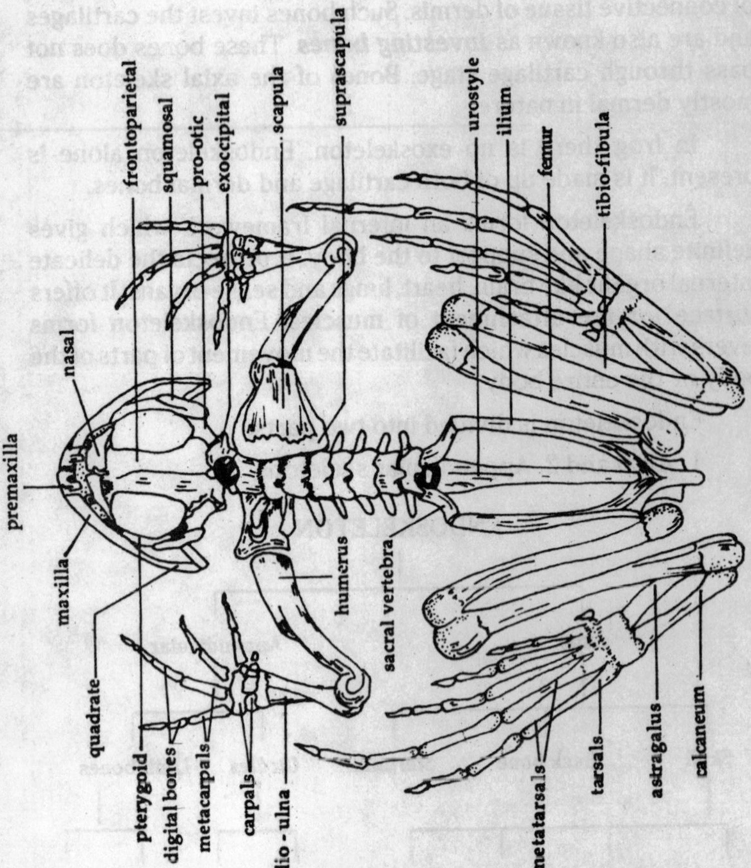
premaxilla
maxilla
quadrate
pterygoid
digital bones
metacarpals
carpals
radio - ulna
nasal
frontoparietal
squamosal
pro-otic
exoccipital
scapula
suprascapula
humerus
sacral vertebra
urostyle
ilium
femur
tibio-fibula
metatarsals
tarsals
astragalus
calcaneum

Skeleton

AXIAL SKELETON

All skeletal parts lying along the median longitudinal axis of the body constitute axial skeleton. In frog it includes ***skull, vertebral column*** and ***sternum***.

SKULL

Skull is the endoskeleton of head. Skull of frog is more or less depressed and triangular in shape. It is described as ***dicondylic*** due to the presence of two occipital condyles. The skull is completely cartilaginous in tadpole stage but becomes mostly bony in the adult frog. The skull consists of 1) cranium 2) sense capsules 3) jaws 4) hyoid apparatus.

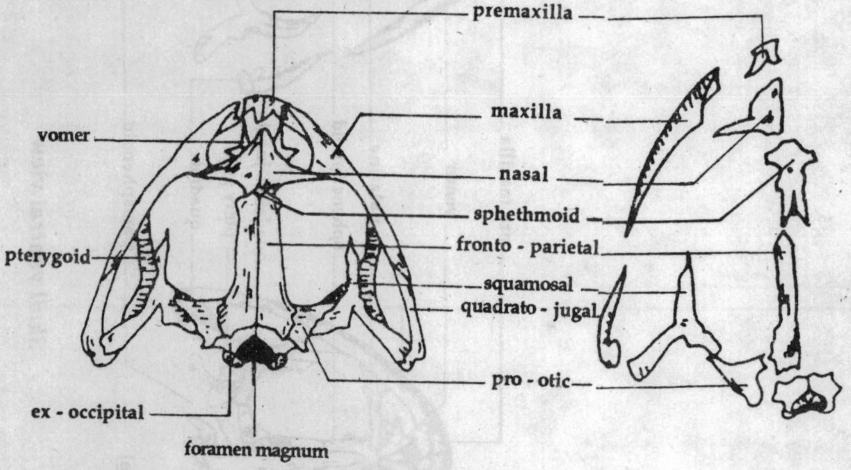

Skull dorsal view

CRANIUM or BRAIN BOX

Cranium is the central part of the skull which encloses a cavity called ***cranial cavity***. The brain is lodged in this cavity. So the cranium is called ***brain box***. Cranium of frog is an oblong, rectangular box of smaller size. The large opening present at the posterior extremity of the cranium is called ***foramen magnum***. The spinal cord comes out through the foramen magnum. The cranium has some foraminae for the passage of nerves and blood vessels. The cranium of frog is made of 6 bones. They are ***2 exoccipitals, 2 fronto parietals, 1 sphenethmoid*** and ***1 parasphenoid***.

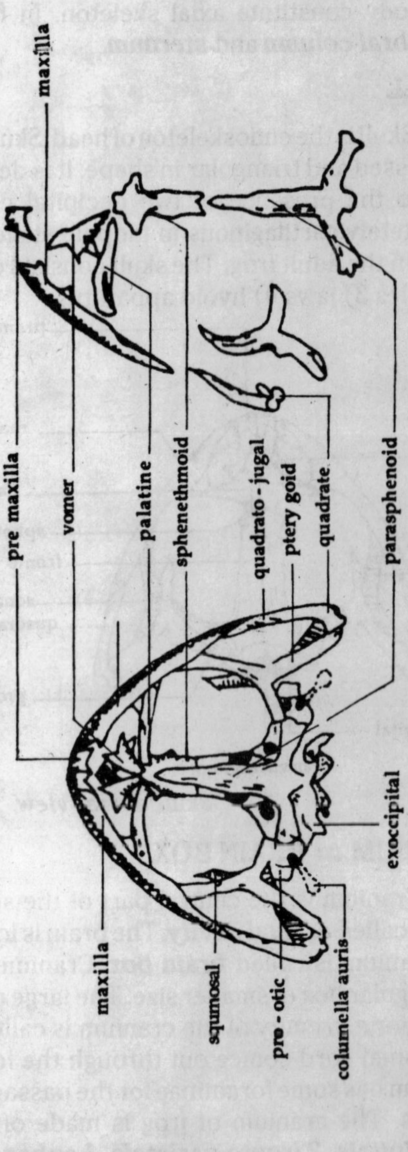

Skull ventral view

The sphenethmoid and exoccipitals are replacing bones whereas the frontoparietals and parasphenoid are membrane bones. The cranium is divisible into occipital, fronto-parietal and Ethmoid regions.

Occipital region

The posterior region of the cranium is called ***occipital region.*** It is formed by the two ***exoccipitals*** situated one on either sie of the foramen magnum. ***Basi and supra occipitals are absent in frog.*** Each exoccipital is produced behind into a knob like elevation called ***occipital condyle.***

The two occipital condyles fit into the articular facets of atlas. This articulation facilitates the up and down movement of the skull on back bone. There is a foramen on each exoccipital just outside the condyle for the passage of the vagus nerve. It is called ***cranial*** or ***jugular foramen.***

Fronto-parietal region.

The middle region of the cranium is called ***frontoparietal region.*** It is formed by two ***fronto-parietals*** and a ***parasphenoid.*** Each frontoparietal is a compound bone formed by the fusion of an anterior frontal and a posterior parietal. The frontals and parietals are distinct in the tadpole, but are fused in the adult. These are flat long bones forming the roof of cranium. The two frontoparietals are united by a suture in the middle line but diverge anteriorly. The floor of the cranium is formed by an inverted 'T' shaped or dagger shaped ***parasphenoid.*** The long arm or blade of parasphenoid extends below the sphenethmoid. The lateral arms extend to the ventral side of auditory capsule.

Ethmoid region

The anterior region of the cranium is called ***ethmoid region.*** It is surrounded by a diamond-shaped bone called the ***sphenethmoid.*** The sphenethmoid is overlapped partly by the diverging fronto-parietals and partly by the nasals on the dorsal side. Hence the sphenethmoid appears as a diamond shaped bone dorsally.

SENSE CAPSULES

Sense capsules are of 3 types:

1. ***auditory capsules.***
2. ***olfactory capsules,*** and
3. ***optic capsules***

Auditory Capsules

There are a pair of auditory capsules situated one on either side of the exoccipital in the posterior end of the cranium. Each auditory capsule is formed by a single saddle shaped cartilage bone called ***prootic***. Each prootic encloses the membranous labyrinth. The prootic on its outersurface bears a membrane covered aperture called ***fenestra ovalis*** into which the proximal part of columella auris fits. ***The columella auris is a modified hyomandibular***. It conducts sound vibrations from the tympanum to the internal ear.

Olfactory capsules

A pair of olfactory capsules are present infront of the cranium which lodge the organs of sense of smell. The dorsal surface or roof of the olfactory capsules is formed by thin and triangular membrane bones called ***nasals***. The two nasals are united in the middle, but diverge posteriorly to meet the frontoparietals. The ventral surface of olfactory capsules is formed by two irregular ***vomers*** bearing ***vomerine teeth***. Both the nasals and vomers are membrane bones.

The external nostrils are supported by two irregular membrane bones called ***septomaxillaries***. The outer margin of nasals together with septomaxillaries from the boundary of external nostrils.

Optic capsules

There are a pair of optic capsules situated one on either side of the middle and anterior parts of cranium. There are no special bones forming the orbits. As there is no interobital septum, the skull of frog is also called ***platibasic***. The optic capsules contain eyes.

UPPER JAW

Upper jaw of frog is in the form of an arch forming the margin of the skull. It consists of two identical halves which are united anteriorly and diverge posteriorly.

Each half of the upper jaw consists of an anterior ***premaxilla,*** middle ***maxilla*** and posterior ***quadratojugal and quadrate.*** All these bones except quadrate, are membrane bones. The premaxillae of both the sides are united in the middle at the anterior end. The premaxilla is followed by a long maxilla which forms the sides of the jaw on each side. Both premaxillae and maxillae bear teeth on the ventral surface. They are called ***maxillary teeth***. The maxilla is followed by the comma-shaped ***quadratojugal*** and is followed posteriorly by cartilaginous Quadrate. The quadratojugal does not bear any teeth.

Each half of the upper jaw is connected with the cranium on each side by three bones viz., ***palatine, pterygoid*** and ***squamosal***. The two palatines are slender, ribbon shaped, transverse bones situated ventrally one on either side and extends between the maxilla and anterior end of sphenethmoid.

The two pterygoids are triradiate bones (Y-shaped bones) extending between the maxilla, quadratojugal and prootic, one on each side posterio-ventrally. Anterior ray of pterygoid is connected with the maxilla. Another backward and inward ray is connected with the prootic. The third backward and outward ray is connected with the Quadrate.

The two squamosals are 'T' shaped bones present on the posterio-dorsal side. The stem of the squamosal is connected with the quadrate while the posterior limb is connected with the prootic. The anterior limb projects into the orbit and remains free.

The upper jaw is firmly fused with the cranium. So the jaw suspension in frog is described as ***autostylic***. Upper jaw is a derivative of ***palatoquadrate cartilage***.

LOWER JAW

Lower jaw or mandible of frog is a derivative of Meckel's cartilage. It is in the form of an arch on the lower side of the skull. The lower jaw consists of two identical halves or rami. The two rami are joined infront by an elastic ligament. Each ramus consists of three bones, viz., ***mentomeckelian, dentary*** and ***angulosplenial***. The mentomeckelian is cartiloginous and the other two bones are membrane.

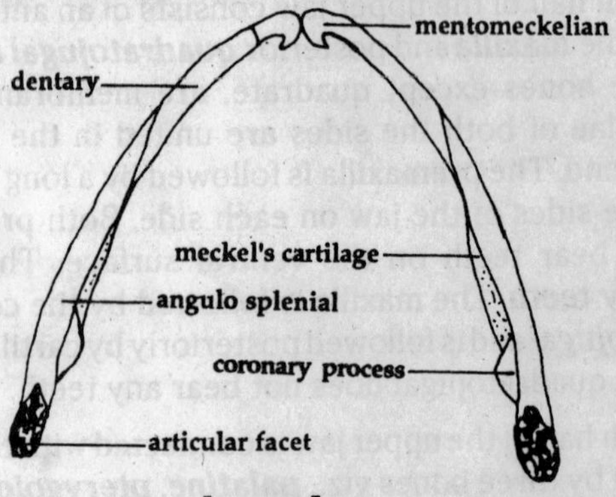

Lower Jaw

The mentomeckelians are joined in the middle by a ligament. The angulo-splenial is a sword shaped bone which invests inner posterior part of Meckel's cartilage. Dentary invests the outer and anterior part of Meckel's cartilage. The angulo-splenial bears a ***coronoid process*** just above the articular facet. The lower jaw joins the pterygoid with the coronoid process. The lower jaw is without teeth.

HYOID APPARATUS

The hyoid apparatus is a part of visceral skeleton of frog. It is found on the floor of the bucco pharyngeal cavity. it is a remnant of the skeleton which supports the gills (branchial skeleton) of the tadpole during development. Hyoid apparatus consists of three parts.

1. ***Body of hyoid.***
2. ***Anterior cornua***
3. ***Posterior coruna.***

The body of hyoid is a broad, thin, flat shield shaped cartilaginous plate. It bears a pair of ***alar processes*** on its anterolateral sides. It offers surface for the attachment of ***sternohyal*** and ***petrohyals.***

The body of hyoid bears a pair of ***anterior cornua***. They arise from the centre of its anterior margin. The anterior cornua are long and slender cartilaginous processes. They curve backwards and finally join the auditory capsule.

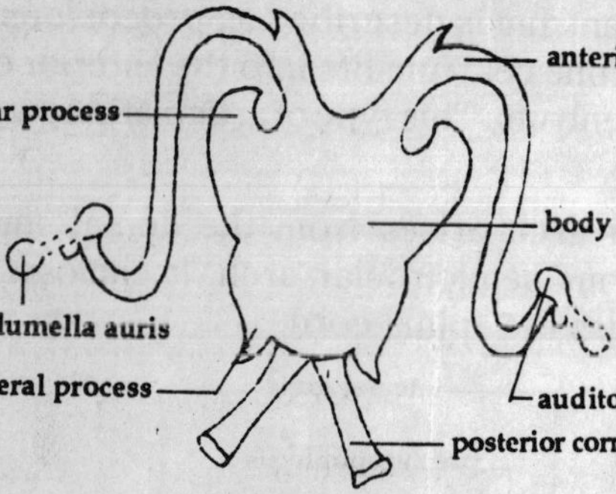

Hyoid apparatus

Posteriorly the body of hyoid bears a pair of short and stout backwardly directed processes called ***posterior cornua***. They extend on either side of the laryngo-tracheal chamber giving support to it. The posterior cornua are replacing bones.

VERTEBRAL COLUMN (BACK BONE)

Vertebral column (back bone) is the part of axial skeleton lying along the middorsal surface of the trunk. It is remarkable for its extreme shortness due to the absence of neck and tail. It consists of ring like vertebrae which are joined by elastic ligaments to facilitate limited movement.The back bone of frog consists of 9 vertebrae and a long rod-like ***urostyle***. There are no cervical and caudal vertebrae due to the absence of neck and tail. The 2–7 vertebrae have similar characters which are described as typical vertebrae. Ist, 8th and 9th vertebrae are peculiar and differ with the typical vertebrae.

Typical vertebra

Each typical vertebra is a flat ring shaped structure. It consists of five parts viz.,

1. ***Centrum***
2. ***Neural arch***
3. ***Neural spine***
4. ***Zygapophyses***
5. ***Transverse processes***

Centrum is a short solid cylindrical bony part found on the ventral side of the vertebra. It is concave infront and convex behind. Such a centrum is described as ***procoelous***. The posterior convexity of one centrum fits into the anterior concavity of the succeeding centrum. This type of articulation gives flexibility to the back bone.

The neural arch arises from the dorsal surface of the centrum and forms semicircular arch. It encloses the neural canal which lodges the spinal cord.

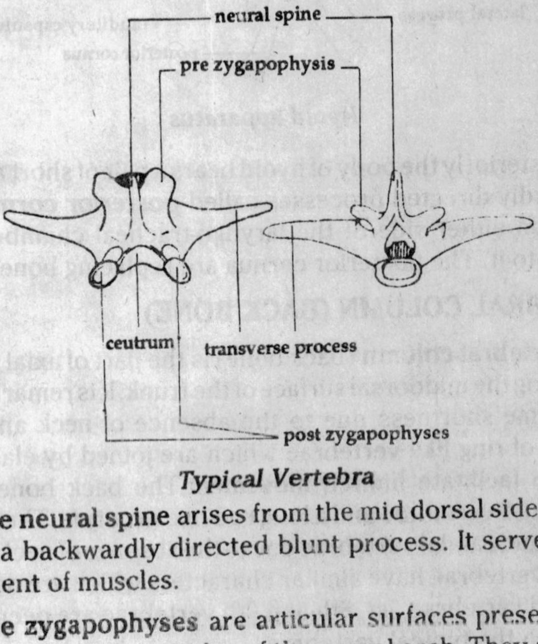

Typical Vertebra

The neural spine arises from the mid dorsal side of neural arch as a backwardly directed blunt process. It serves for the attachment of muscles.

The zygapophyses are articular surfaces present on the anterior and posterior sides of the neural arch. They are called ***Pre Zygapophyses*** and ***Post Zygapophyses*** respectively. The prezygapophyses are directed upwards and inwards, whereas the ***postzygapophyses*** face downwards and outwards. The postzygapophyses of one vertebra overlap the prezygapophyses of the next vertebra forming gliding joints.

The two transverse processes are long and laterally directed bony rods, situated one on either side of neural arch. They offer surface for the insertion of muscles and support the dorsal body wall.

The openings present on the lateral sides of the vertebral column in between the successive vertebrae are called ***inter vertebral foramina***. The spinal nerves emerge out through these foramina.

ATLAS

Atlas is the first or cervical vertebra which articulates with the skull. It supports the skull and bears the entire weight of the skull. It is highly modified to serve this purpose.

It is a ring like vertebra and centrum is highly reduced.

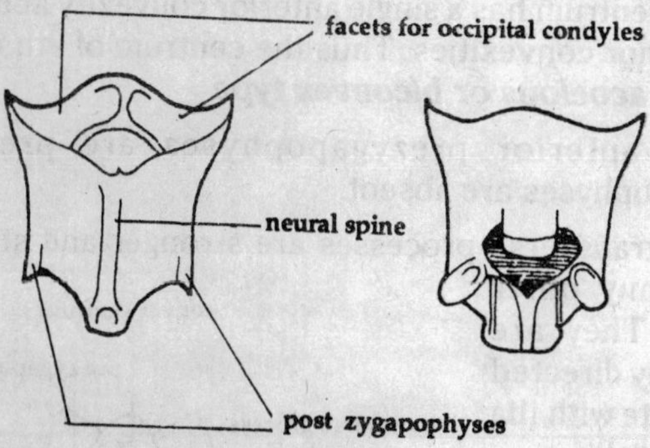

Atlas

It is characterized by the absence of transverse processes and prezygapophyses. The postzygapophyses and neural spine are reduced.

The centrum bears two concave articular facets at its anterior face to receive the two occipital condyles of the skull.

8TH VERTEBRA

The 8th vertebra of frog has a centrum which is concave infront and behind. Such a vertebra with biconcave centrum is termed as ***amphicoelous vertebra***.

It also bears a pair of thin pointed transverse process.

The 8th vertebra is also known as ***lumbar vertebra***. The anterior concavity receives the posterior convexity of the preceeding 7th vertebra, and the posterior concavity accommodates the anterior convexity of 9th vertebra. It is similar to typical vertebra in rest of the characters.

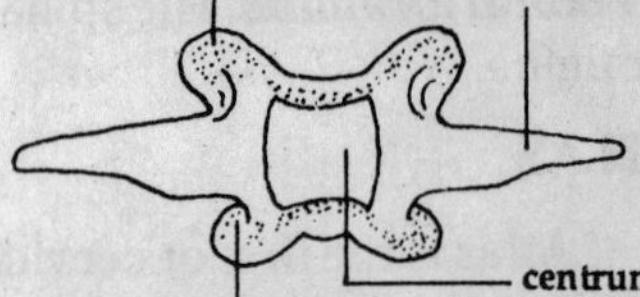

8th Vertebra

9TH VERTEBRA

The 9th vertebra of frog is called ***sacral vertebra***, as it forms a sacrum with its transverse processes which articulate with ilia of pelvic girdle. It provides firmness to the hinder region of trunk.

The centrum has a single anterior convexity and two knob-like posterior convexities. Thus the centrum of 9th vertebra is termed as ***acoelous or biconvex type***.

The anterior prezygapophyses are present and postzygapophyses are absent.

The transverse processes are stronger and stouter than those of any other vertebra. They are backwardly directed to articulate with ilia of pelvic girdle.

The two posterior convexities of the 9th vertebra fit into the two depressions of urostyle.

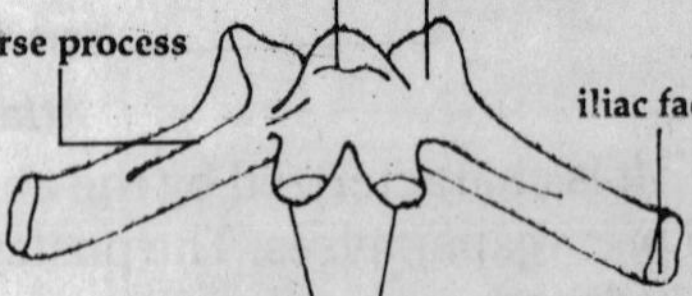

9th Vertebra

UROSTYLE

The urostyle is a rod like structure and is believed to be formed by the fusion of caudal vertebrae. According to modern view, it is formed by the ossification of a single rod-like cartilage.

The urostyle has a rod-shaped centrum. It is broad anteriorly and tapers posteriorly.

Urostyle lies in the space between the two ilia of pelvic girdle. It is characterised by the absence of transverse processes, zygapophyses and neural spine.

It has a ***dorsal crest*** which encloses a minute neural canal for the passage of filum terminate of spinal cord.

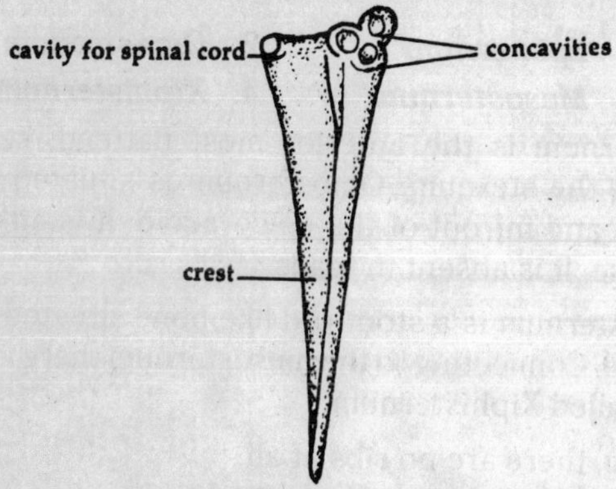

Urostyle

The neural canal of urostyle bears dorsally a foramen on either side or one side near its anterior end through which the last pair of spinal nerves come out.

It bears two concavities anteriorly into which the two posterior condyles of 9th vertebra fit.

The length of the urostyle is more or less equal to the length of the rest of the back bone.

FUNCTIONS OF VERTEBRAL COLUMN

Vertebral column protects the spinal cord.

It forms a sort of beam along the middorsal surface and offers good support to the trunk.

It is directly attached with the pelvic girdle to provide a firm base for the articulation of hind limbs.

Back bone allows bodily flexions.

It offers surface for the attachment of muscles.

STERNUM

The sternum or breast bone is a part of axial skeleton. ***Sternum is first formed in Amphibia***. However, it is absent in caecilians.

The sternum lies along the midventral line. It binds the two halves of pectoral girdle anteriorly and posteriorly.

The sternum consists of 4 parts.

1. ***Episternum*** 2. ***Omosternum***
3. ***Mesosternum*** 4. ***Xiphisternum***

Episternum is the anterior most flattend, semi-circular cartilage of the sternum. Omosternum is situated behind the episternum and infront of the epicoracoid. It is an inverted Y-shaped bone. It is absent in toads.

Mesosternum is a stout rod-like bone situated behind the epicoracoid. Connecting to the mesosternum there is a large flat cartilage called Xiphisternum.

In frog, there are no ribs at all.

Functions

Sternum protects the heart and lungs.

It gives support to the ventral part of the thoracic region.

Sternum provides place for the insertion of sternohylals at one end.

APPENDICULAR SKELETON

All skeletal parts lying on either side of the long axis of the body which are appended to the axial skeleton, constitute the appendicular skeleton. It consists of bones of girdles and limbs. There are two girdles viz., ***pectoral*** and ***pelvic***.

PECTORAL GIRDLE

The pectoral girdle or the shoulder girdle is in the form of an inverted arch. It is enclosed in the mucles of the anterior part of the trunk encircling the back bone from 2–4 vertebrae.

The pectoral girdle consists of two similar halves which are united midventrally by sternum but separated above dorsally. Each half of the pectoral girdle is made of 1. ***suprascapula*** 2. ***scapula***, 3. ***coracoid*** 4. ***calvicle*** 5. ***precoracoid***, 6. ***Epicoracoid***.

The suprascapula, precoracoid and epicoracoid are cartilages and the rest are bones.

Each half of the pectoral girdle consists of dorsally, a broad calcified cartilage called ***suprascapula***. It arches over the dorso-lateral sides of backbone. The inner free edge of suprascapula bears a marginal strip of ***hyaline cartilage***. The outer face of suprascapula is joined with the outer face of scapula.

The scapula is a short, stout and flat bone lying above the glenoid cavity. The inner end of scapula is attached to the coracoid on the ventral side.

The glenoid cavity is a depression formed partly by the scapula and partly by the coracoid in each half of the pectoral girdle. The head of the humerus fits into the glenoid cavity forming ***a ball and socket joint***. Thus the fore limb articulates with the pectoral girdle.

The coracoid is a stout dumble-shaped bone. It is found on the ventral side of glenoid cavity in postero-lateral position. Coracoid is attached to the scapula. The coracoids of both halves of pectoral girdle are joined together by a cartilaginous epicoracoid in the midventral surface.

The calvicle or collar bone is a long slender rod like bone lying transversely ventral and anterior to glenoid cavity in each half. A strip of cartilage called ***precoracoid*** is found beneath each calvicle.

The large gap found in each half between the clavicle and coracoid is called ***coracoid fenestra***. (Include here sternum which is described earlier).

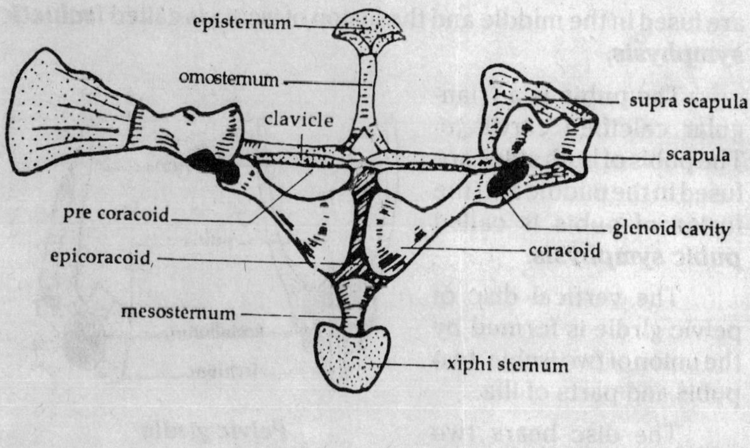

Pectoral girdle

Functions

Pectoral girdle protects the internal organs such as lungs and heart.

It gives proper shape and strength to the body.

If offers surface for the attachment of brachial and shoulder muscles.

It binds the bones of fore limb with the back bone.

It reduces the intensity of mechanichal shocks while leaping over the ground.

PELVIC GIRDLE

The pelvic girdle of frog is a long 'V' shaped well built skeletal part. It is present at the posterior part of the trunk.

The pelvic girdle consists of two similar halves called ***os-innominata***. Each half or os-innominatum consists of 3 bones viz. 1. ***Ilium*** 2. ***Ischium*** and 3. ***Pubis***.

The ilia are well developed where as pubes and ischia are reduced.

Each **ilium** is a long elongated bone forming major part of each os-innominatum. Anteriorly the two stout processes of ilia articulate with the transverse processes of the 9th vertebra. The ilium bears a vertical ridge called ***iliac crest*** on the dorsal side.

The **ischium** is a reduced triangular bone. The two ischia are fused in the middle and the fusion of ischia is called ***ischiatic symphysis***.

The **pubis** is a triangular calcified cartilage. The pubis of both sides are fused in the middle and the fusion of pubis is called ***pubic symphysis***.

The vertical disc of pelvic girdle is formed by the union of two ischia, two pubis and parts of ilia.

The disc bears two

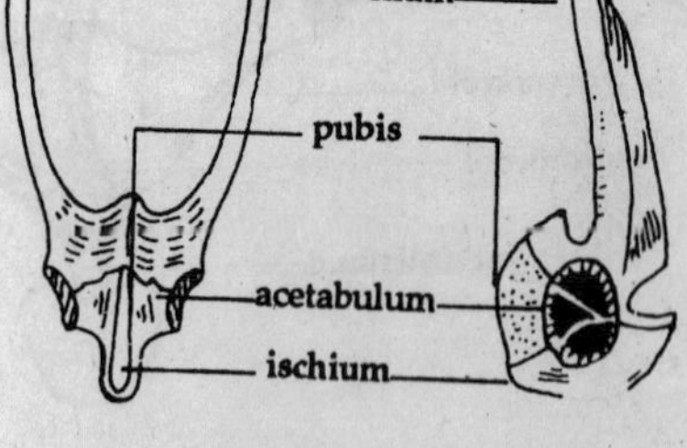

Pelvic girdle

lateral depressions one on either side. They are called ***acetabulae***.

The head of femur fits into the acetabulum forming the ball and socket joint.

Functions

The pelvic girdle gives adequate support to the back bone and firmness to the posterior end of trunk.

It articulates the hind limbs with backbone.

The elongated ilia of pelvic girdle help in the absorption of mechanical shocks during leaping.

It provides surface for the attachment of hip muscles.

LIMB SKELETON

The frog has two pairs of pentadactyle limbs called ***fore limbs*** and ***hind limbs***. The bones of the hind limb have become elongated. The bones of the fore limb have become short and stout to give support to the body during leaping.

Fore Limb Skeleton

The fore limb of frog consists of an ***upper arm (brachium) fore arm (anti brachium)*** and ***hand. (Manus)***. The hand (manus) is again divisible into wrist, ***palm*** and ***fingers***. The bones present in these five parts of the fore limb are as follows :

Limb part	***Bone(s)***
1. Upper arm	***Humerus***
2. Fore arm	***Radio-ulna***
3. Wrist	***Carpals***
4. Palm	***Metacarpals***
5. Finger	***Phalanges***

Humerus

The bone of the upper arm is called ***humerus***. It is a short, stout and slightly curved bone.

The proximal head of humerus is capped by a ***calcified cartilage***. The head of the humerus fits into the glenoid cavity of pectoral girdle. The distal end of humerus bears a spherical ***capitulum with trochlea***.

The shaft of the humerus near its head bears a ***deltoid ridge***. It provides surface for the attachment of muscles. The deltoid ridge is the distinguished feature of humerus.

Radio-ulna

The bone of the fore arm is ***radio-ulna***. It is a compound bone formed by the union of radius and ulna. The two bones are distinct in the larval stage. The fusion is indicated by the presence of a groove near its distal end.

The proximal end of radio-ulna has a a concavity to receive the capitulum of the distal end of humerus. This articulation forms an ***elbow joint***.

The proximal end of ulna bears an ***olecranon process*** behind the elbow joint.

The distal end of radio-ulna is broad and bears two unequal articular facets.

Carpals

The bones of the wrist are called ***carpals***. There are six carpals arranged in two rows of 3 each.

The proximal row of carpals are the ***radiale, centrale or intermediale and ulnare.***

The distal row of carpals are called ***capitohamatum, trapezoid, trapezium.***

According to one view, the 2nd and 3rd distal carpals are supposed to have fused to form a compound carpal.

Metacarpals

The bones of the palm are called ***metacarpals.***

There are 5 metacarpals in the palm in one row.

The first metacarpal is rudimentary and hidden underneath the skin. Some authors treated the inner reduced metacarpal as the ***pollex***. The remaining four meta-carpals are long and bear phallanges.

Phallanges

The hand of frog has four fingers (digits) only. The digit corresponding to our thumb is absent. The first two digits have

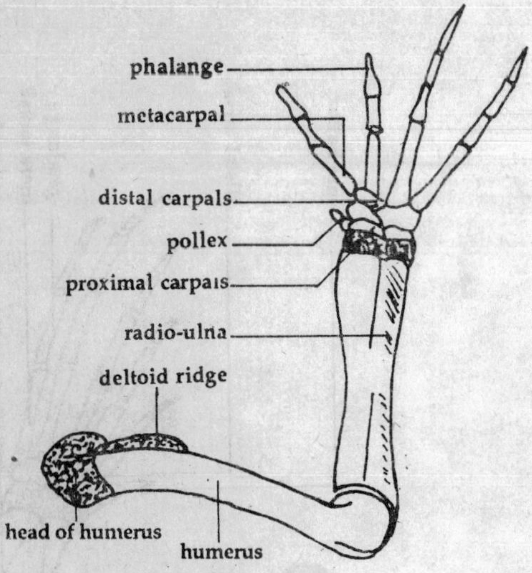

Fore limb skeleton

2 phallanges each. The 3rd and 4th fingers have 3 phallanges each. The phallanges of the fore limb are represented by a phallangeal formula 1,2,2,3,3.

Hind Limb Skeleton

The hind limb consists of ***thigh, shank and foot***. Foot is further divided into ***ankle, instep and digits or toes***. The bones found in the different parts of the hind limb are as follows:

Limb part	***Bone(s)***
1. Thigh	***Femur***
2. Shank	***Tibio-fibula***
3. Ankle	***Tarsals***
4. Instep	***Metatarsals***
5. Toes	***Phalanges***

Femur

Thigh bone or femur is a long slender curved bone. The proximal end or head of femur fits into the acetabulum of pelvic

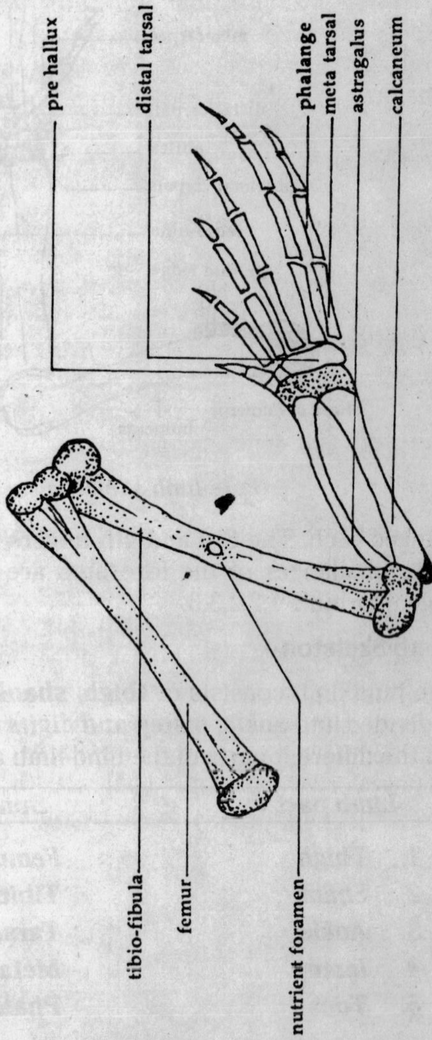

Hind limb skeleton

girdle to form a ball and socket joint. The distal end of femur is flattened to articulate with the distal end of tibio-fibula.

Tibio-fubula

It consists of inner tibia and outer fibula. The fusion is indicated by the presence of grooves at both ends. The **shaft** bears a ***nutrient foramen*** through which blood vessels enter the bone. The proximal end of tibio-fibula articulates with the distal end of femur forming ***knee joint***. The distal end of tibiofibula articulates with the proximal row of tarsals.

Tarsals

The bones of the ankle are called ***tarsals***. There are four tarsals arranged in two rows. The proximal row of two tarsals are elongated and united at both ends leaving a wide gap in the middle. The elongated inner tibiale and outer fibulare are called ***astragalus and calcaneum*** respectively. The elongated tarsals make the ankle long to facilitate leaping. The distal row of two tarsals have no distinct names in frog.

Metatarsals

The bones of the instep or sole are called ***metatarsals.*** There are 5 metatarsals in the instep of frog. There is an additional or supplementary toe called ***Prehallux***. It is hidden underneath the skin. The five metatarsals bear five toes. The first toe is called ***hallux***.

Phalanges

The bones toes are called ***Phalanges***. The first two toes bear two phalanges each. The 3rd bears 3, 4th bears 4 and the 5th bears 3 phallenges. The phalangeal formula is 1,2,2,3,4,3.

5

Digestive System

The digestive system includes various organs concerned with ingestion, digestion, absorption and elimination of undigested waste. As the above mentioned processes are also concerned with nutrition, this system is also called ***nutritive system***. Based on the food habits, the animals are divided into three categorics 1. herbivores 2. carnivores 3. omnivores. Most of the animals ingest food materials which include proteins, fats and carbohydrates, vitamins and mineral salts.

The alimentary canal of frog is longer than the body. Therefore it is found in the coiled condition. It is generally long in herbivores and short in carnivores. The parts of alimentary canal are variously modified to suit to feeding habits of various animals.

The digestive system of frog mainly consists of ***alimentary canal*** and ***digestive glands***. The food is digested and absorbed in the alimentary canal, while the digestive glands secrete certain enzymes which help in the digestion of ingested food.

ALIMENTARY CANAL

The alimentary canal is long coiled tube of varying diameter which extends from mouth to cloaca. It consists of a ***bucco-pharyngeal cavity, oesophagus, stomach, small intestine, rectum*** and ***cloaca***.

BUCCO-PHARYNGEAL CAVITY

The mouth is situated at the anterior end of the head. It leads into a wide and broad cavity called ***bucco pharyngeal cavity***. It is bounded by the upper and lower jaws. The tip of lower jaw bears three projections called ***prelingual tubercles***. They fit into the ***sub rostral fossae*** of the upper jaw and the mouth is tightly closed. The margin of upper jaw has a groove called, the ***sulcus marginalis***. The lower jaw fits into this groove to close the mouth air tight. The upper jaw bears a row of small, uniform and hook-like pointed ***maxillary teeth***.

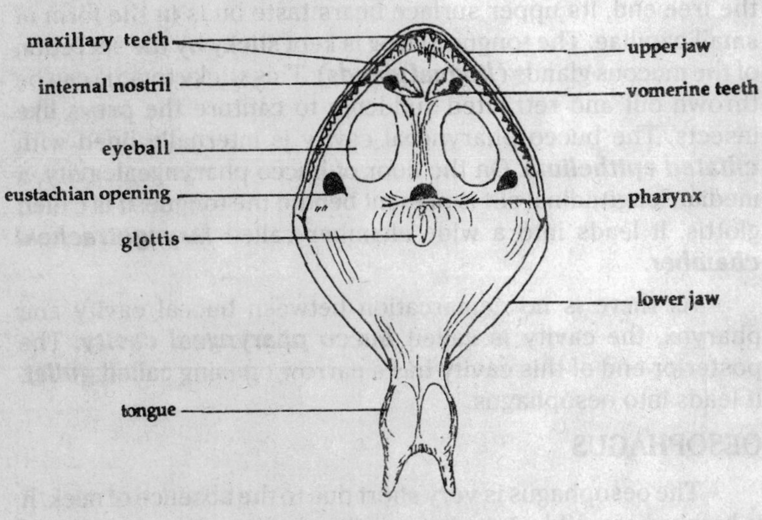

Bucco-pharyngeal Cavity

The lower jaw lacks teeth. Besides the maxillary teeth, three are two sets of ***vomerine teeth*** on the roof of the buccal cavity, one set behind each internal nostril. The teeth are not useful for chewing the food, but they prevent the escape of the live prey. They are curved backwards and directly attached to the jaw bones instead of being set in sockets. Such teeth are called ***acrodont***. The teeth are all similar in shape and size, therefor they are ***homodont*** or ***isodont***.. As the teeth are replaced several times during the life of frog, they are described as ***polyphyodont***. As the teeth are attached to the lateral sides of jaw bones, they are also known as ***pleurodont teeth***.

On the roof of the buccal cavity and infront of the vomerine teeth, there are 2 openings, the ***internal nostrils***. They communicate to out side through a pair of external nostrils. A pair of openings of eustachian tubes are situated at the sides of posterior part of the bucco pharyngeal cavity. A pair of eye balls are found as bulges on the roof of the bucco pharyngeal cavity. They also help in preventing the escape of the prey.

A large, thick, sticky and, protrusible tongue is present on the floor of the bucco pharyngeal cavity. The tongue is peculiarly attached infront and free behind. The tongue is slightly bifid at the free end. Its upper surface bears taste buds in the form of small papillae. The tongue of frog is kept sticky by the secretion of the mucous glands (***lingual glands***). This sticky tongue can be thrown out and retracted suddenly to capture the preys like insects. The bucco pharyngeal cavity is internally lined with ***ciliated epithelium***. On the floor of bucco pharyngealcavity, a median longitudinal slit is present behind the tongue. It is called glottis. It leads into a wide chamber called ***laryngotracheal chamber***.

As there is no demarcation between buccal cavity and pharynx, the cavity is called ***bucco pharyngeal cavity***. The posterior end of this cavity has a narrow opening called ***gullet***. It leads into oesophagus.

OESOPHAGUS

The oesophagus is very short due to the absence of neck. It is highly distensible as its lining is thrown into a large number of longitudinal folds called ***rugae***. These folds allow sufficient expansion of the oesophagus during the passage of food through it into the stomach. The inner wall of the oesophagus has ***goblet cells*** and it opens into stomach.

STOMACH

It is the muscular sac-like part of the alimentary canal situated behind the oesophagus. It is suspended by a mesentery into the left side of the body cavity.

The stomach is distinguished into an anterior wide ***cardiac stomach*** and a posterior short and narrow ***pyloric stomach***. The internal lining of the stomach forms numerous longitudinal folds

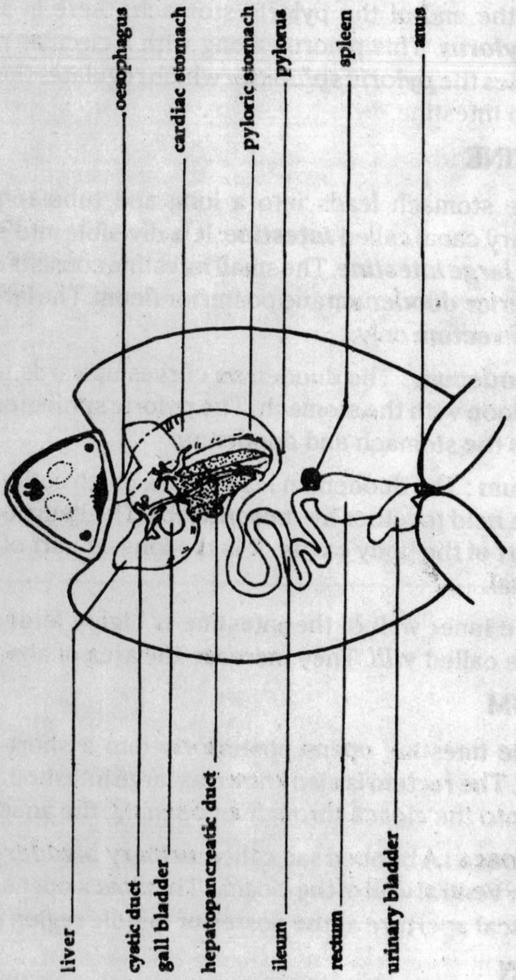

Digestive System

called ***rugae***. These folds allow the expansion of the stomach. The tubular gastric glands present in the mucous layer of stomach secrete ***gastric juice***, which contains pepsinogen and HCl.

At the end of the pyloric stomach, there is an aperture called ***pylorus***. This pylorus along with a circular ring muscle constitutes the ***pyloric sphincter*** which regulates the passage of food into intestine.

INTESTINE

The stomach leads into a long and tubular part of the alimentary canal called ***intestine***. It is divisible into ***small intestine*** and ***large intestine***. The small intestine consists of two parts viz., anterior ***duodenum*** and posterior ***ileum***. The large intestine includes ***rectum*** only.

Duodenum : The duodenum curves upwards to form a 'U' shaped loop with the stomach. The pyloric sphincter is situated between the stomach and duodenum.

Ileum : The duodenum leads into a much coiled ileum. The coils are held together by ***mesenteries***. The ileum occupies the right part of the body cavity. It is the longest part of the alimentary canal.

The inner wall of the intestine is highly folded and these folds are called ***villi***. They increase the area of absorption.

RECTUM

The intestine opens posteriorly into a short thin walled ***rectum***. The rectum is also known as large intestine. The rectum opens into the cloaca through an opening, the anus.

Cloaca : A bilobed sac called ***urinary bladder*** is produced from the ventral wall of the cloaca. The cloaca opens out through the cloacal aperture at the posterior middle region of the trunk.

SPLEEN

Spleen is found attached to the mesentery at the junction of the ileum and rectum by a fold of peritoneum. Although it is closely associated with the alimentary canal, it is not concerned with the process of digestion.

HISTOLOGY OF THE ALIMENTARY CANAL

The wall of the alimentary canal is made up of four distinct layers. Counting from outside, these layers are:

1. ***the visceral peritoneum of serous layer,***
2. ***the muscular layer,***
3. ***the submucosa,*** and
4. ***the mucosa***

Serous layer is the outermost layer. It is composed of visceral peritoneum called ***serosa.***

Muscular layer is composed of outer longitudinal muscles and inner circular muscles. The circular muscle layer and mucosa in the stomach are thick.

Sub-mucosa is a thin layer composed of loose connective tissue, elastic fibres, flood and lymph vessels, nerve cells and nerve fibres.

Mucosa is the innermost layer composed of connective tissue, blood vessels, lymph vessels, nerves, and a basement membrane supporting a layer of columnar epithelial cells. Some of them are known as ***goblet cells***, which secrete mucous. The mucosa in intestinal region is thrown into finger like folds called ***villi***. The villus is a unit of absorption containing a lymph vessel called ***lacteal*** and a net work of capillaries. The area of absorption of digested food is increased by villi.

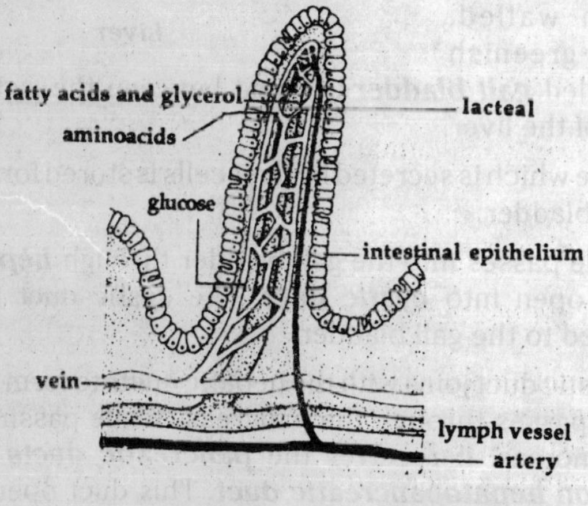

Section of intestine (Villus enlarged)

DIGESTIVE GLANDS

There are numerous glands situated in the mucous membrane of the alimentary canal. Two large digestive glands are also found out side the alimentary canal of frog. They pour their secretions into the lumen of the intestine through their ducts.

LIVER

- It is one of the most important vital organs and is the largest gland found in the body of the frog.
- It is the dark reddish brown glands situated in the anterior region of the body cavity.
- It consists of two unequal lobes, the right and the left. The left lobe is the largest and covers greater part of the stomach. This is subdivided into two lobes.
- A thin walled, round greenish sac called ***gall bladder*** is found between the right and left lobes of the liver.

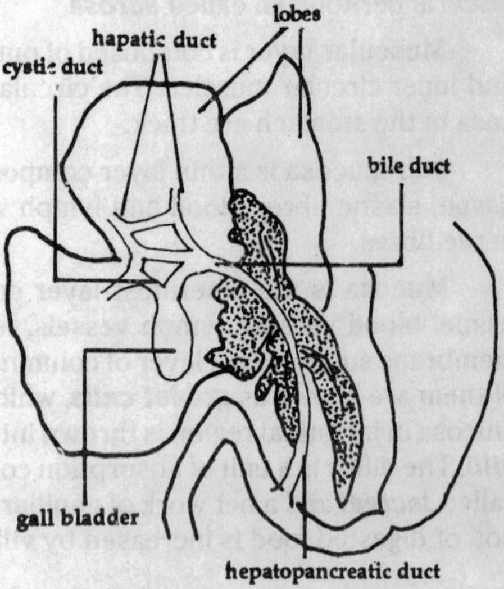

Liver

- The bile which is secreted by liver cells is stored for some time in gall bladder.
- The bile passes into the gall bladder through ***hepatic ducts*** which open into ***cystic duct***. The cystic duct is directly attached to the gall bladder.
- The cystic duct joins with the hepatic ducts to form a ***bile duct*** which passes through the pancreas. While passing through the pancreas, it receives the ***pancreatic ducts*** to form a ***common hepatopancreatic duct.*** This duct opens into the duodenum.

FUNCTIONS OF THE LIVER

• The liver secretes a greenish alkaline fluid called bile. Bile contains bile salts, and bile pigments. Enzymes are absent in bile. The sodium bicarbonate present in bile reduces the acidity of the food in the intestine and provides alkaline medium.

• The bile salts viz, ***taurocholate*** and ***glycocholate of sodium and potassium*** help in the digestion of the fats by emulsifying them.

• It stores excess sugar in the form of glycogen (animal starch). The liver converts excess glucose into glycogen (***glycogenesis***). It also converts glycogen into glucose (***glycogenolysis***), whenever the concentration of glucose falls and thus regulates the blood sugar level. Liver also converts other organic substances like proteins and fats into glucose (***gluconeogenesis***).

• The liver promotes ***lipogenesis***. It is a process in which excess carbohydrates and converted into fats.

• It maintains the protein concentration of the blood. The toxic ammonia and excess proteins are converted into urea in liver (***deamination***) which is finally eliminated by kidneys in the form of urine.

• It also removes excretory waste products like bile pigments viz., ***bilirubin and biliverdin***. These are the breakdown products of haemoglobin which are passed out along with faeces.

• It helps in the production of ***fibrinogen*** and ***prothrombin*** which are essential for blood clotting.

• It produces red blood corpuscles during embryonic stage (***huemopoietic***).

• It also produces ***heparin***, an intravascular anticoagulant which prevents the clotting of blood in blood vessels.

• It reabsorbs and stores copper and iron. It also produces and stores vitamin A.

• Liver plays an important role in phagocytosis of foreign organisms like bacteria with the help of ***kupffer cells*** i.e. the reticuloendothelial cells.

• Liver is an organ of detoxification i.e. it eliminates toxic substances.

• Liver also functions as a ***haemolytic organ*** because it destroys old and worm out RBC and they are eliminated in the form of bile pigments.

PANCREAS

• It is second largest, flattened yellowish brown gland with exocrine and endocrine functions.

• It is held by a mesentery between the stomach and duodenum. It is traversed by bile duct into which the pancreatic ducts also open and form ***hepato-pancreatic duct***. It opens into duodenum.

• The pancreas is composed of several lobes an each lobe in its turn consists of several lobules made up of ***acinar cells***. A connective tissue having blood vessels and nerves is present between the lobules.

• The pancreas consists of exocrine and endocrine parts. The exocrine part produces ***pancreatic juice*** which contains sodium bicarbonate and three enzymes viz., ***lipase or steapsin, amylopsin and trypsinogen.***

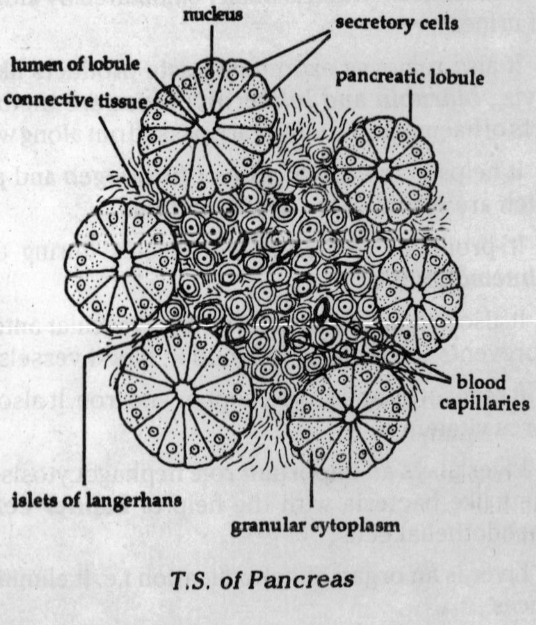

T.S. of Pancreas

• The endocrine part of pancreas consists of several groups of cells called ***islets of Langerhans***. These cells produce two hormones viz., ***glucagon and insulin.***

GASTRIC GLANDS

Gastric Glands : These are the small tubular branched glands present in the mucosa layer of stomach. The gastric glands of cardiac stomach are called ***cardiac glands*** while that of pyloric region are called ***pyloric glands***.

The gastric glands have three types of cells, viz., ***mucous cells*** which produce mucous, ***oxyntic cells*** producing hydrochloric acid and ***zymogen cells*** or ***peptic cells*** producing ***pepsinogen***. The mixture of pepsinogen and hydrochloric acid forms the ***gastric juice***.

INTESTINAL GLANDS

The inner wall of intestine consists of ***crypts of Lieberkuhn*** which secrete intestinal juice called ***succus entericus***. The enzymes like ***erepsin, maltase*** and ***lipase*** are present in the intestinal juice.

PHYSIOLOGY OF NUTRITION

Nutrition is an anabolic process. It is of holozoic type in frog. It involves, digestion, absorption assimilation and egestion.

INGESTION

Frog is a carnivore. It feeds on worms, insects, spiders, snails, smaller frog, etc. Frog captures the prey by the rapid action of sticky tongue. When the frog is nearer to, it opens its mouth and suddenly shoots out its sticky tongue. They prey gets attached to the sticky tongue and is with drawn. Immediately the mouth is closed and the eye balls are depressed. The escape of the living prey is prevented by the maxillary and vomerine teeth. As the teeth are not useful in mastication, frog simply swallows its food. From the bucco-pharyngeal cavity the prey is pushed into the oesophagus.

PROCESS OF DIGESTION

The process of conversion of unabsorbable macromolecules of food (proteins, carbohydrates, fats) into soluble and absorbable micromolecules (aminoacids, glucose and fattyacids and glycerol) by physical and chemical changes is called digestion. The physical changes are brought about by the peristaltic movements of alimentary canal while the chemical changes are brought about by the organic biocatalysts called, ***enzymes.***

$$\text{unabsorbable organic food materials} \xrightarrow[\text{Enzymes}]{H_2O} \text{absorbable simple food materials}$$

The proteins are digested by ***proteolytic enzymes*** or ***proteases***, carbohydrates by ***carbohydrases*** or ***amylases*** and fats by ***lypolytic enzymes or lipases.***

In frog no digestion takes place in the buccopharyngeal cavity as the salivary glands are absent. When the food passes through the oesophagus, the secretions of oesophagus are released into the food. No digestion takes place in oesophagus as the pepsinogen secreted by the oesophagus is inactive.

DIGESTION IN STOMACH

As soon as the food reaches the stomach, it is thoroughly mixed up with ***gastric juice*** secreted by the ***gastric glands*** situated in the internal lining of the stomach. The gastric glands are activated by a hormone called ***gastrin*** produced from the wall of stomach. The gastric juice contains a large amount of water, mucous inactive ***pepsinogen*** and ***dilute hydrochloric acid.*** Inactive pepsinogen is converted into active pepsin on mixing with hydrochloric acid. The acid also kills bacteria, dissolves the inorganic salts and makes the food soft. The enzyme ***pepsin*** acts on proteins and convert them into ***proteoses and peptones.*** It acts most efficiently between the range of *1.5–2* pH. The mucous protects the stomach from the action of the pepsin and hydrochloric acid.

$$\text{Pepsinogen} + HCl \longrightarrow \text{Active pepsin}$$

$$\text{Proteins} + H_2O \xrightarrow{\text{Pepsin}} \text{Proteoses} + \text{Peptones}$$

Thus the food is partly digested in the stomach and is converted into thick, acidic fluid called ***chyme.*** The pylorus opens by this time to allow the chyme to enter the duodenum in small quantities.

DIGESTION IN THE INTESTINE

As the acidic food enters the duodenum, the hydrochloric acid present in food stimulates the wall of duodenum to produce two hormones called secretin and ***cholecystokinin***. Secretin stimulates liver and pancreas to secrete ***bile juice*** and ***pancreatic juices*** respectively. Cholecystokinin constricts the gall bladder to release the bile juice. At the same time another hormone called ***enterocrinin*** stimulates the intestinal glands to secrete intestinal juice called ***succus entericus***. ***Enterocrinin*** is secreted by the wall of small intestine itself. Thus the food in duodenum is mixed with three types of digestive juices. All these act on food to digest it thoroughly.

Bile juice

It is an alkaline fluid secreted by liver. It is temporarily stored in gall bladder and released into the lumen of the duodenum only at the time of digestion. Bile contains no digestive enzymes. But it helps in the emulsification of fats by breaking them into minute droplets. Bile juice also activates the pancreatic lipase. Bile salts are also essential for absorption of digested fats.

Pancreatic juice

It is also an alkaline fluid containing water, inorganic salts and three enzymes namely, ***trypsinogen, exopeptidases, amylopsin (amylase) and lipase (steapsin)***. Trypsinogen is an inactive enzyme while amylopsin and lipase are active enzyme. Inactive trypsinogen is mixed with an enzyme activator called ***enterokinase*** of succus entericus to become an active trypsin. ***Trypsin*** acts on ***proteoses, peptones*** and undigested ***proteins*** changing them into simple proteins called polypeptides. The exopeptidases convert ***polypeptides*** into dipeptides and amino acids. ***Amylopsin*** acts on ***starches*** to convert them into ***maltose*** sugar. ***Steapsin*** acts on emulsified ***fats*** and changes them into ***diglycerides*** and ***monoglycerides***.

$$\text{Trypsinogen} + \xrightarrow{\text{Enterokinase}} \text{Active trypsin}$$

$$\text{Proteins, Proteoses, Peptones} + H_2O$$

$$\xrightarrow{\text{Trypsin}} \text{Polypeptides} \xrightarrow{\text{Exopeptidases}}$$

dipeptides and amino acids

$$\text{Starch} + H_2O \xrightarrow{\text{Amylopsin}} \text{Maltose}$$
(Sugar)

$$\text{Emulsified fats} + H_2O \xrightarrow{\text{Lipase}} \text{Diglycerides}$$

Intestinal Juice

It is also an alkaline fluid which contains enzymes like ***peptidase, lipase, maltase, invertase*** and enzyme activator called ***enterokinase*** (enteropeptidase).

Peptidases include proteolytic enzymes like ***erepsin*** which act on polypeptides changing them into amino acids. ***Lipase*** acts on partly digested fats changing them into ***fatty acids*** and ***glycerol***. ***Maltase*** hydrolyses ***maltose*** sugar to glucose. ***Invertase*** acts on ***sucrose*** to change it to glucose and fructose.

$$\text{Polypeptides} + H_2O \xrightarrow{\text{Erepsin}} \text{Amino acids}$$

$$\text{Partly digested fats} + H_2O \xrightarrow{\text{Lipase}} \text{Fatty acids}$$
$$+ \text{Gycerol (monoglycerides)}$$

$$\text{Maltose} + H_2O \xrightarrow{\text{Maltase}} \text{Glucose}$$

$$\text{Sucrose} + H_2O \xrightarrow{\text{Invertase}} \text{Glucose} + \text{Fructose}$$

Finally the proteins, fats and carbohydrates (starches) are hydrolysed into aminoacids, fatty acids and glycerol and glucose or similar monosaccharide sugars respectively. Thus the digestion of various food stuffs is completed in the intestine. This digested food formed in the small intestine is called ***chyle***.

ABSORPTION

The process by which the digested food is absorbed into blood is called ***absorption***. This absorption takes place either by ***active transport or passive transport***. Absorption mainly takes place in the duodenum and ileum of alimentary canal.

The absorption of food is aided by the presence of finger-like villi which increase the absorptive surface. Each villus is

DIGESTIVE ENZYMES

Digestive Gland	Digestive Juice	Enzyme	Enzyme acts on (substrate)	Enzyme converts food into (end product)
Oesophageal glands	——	Pepsinogen	——	——
Gastric glands	Gastric Juice	Pepsin	Proteins	Peptones, Proteoses
Liver	Bile juice	——	——	Note : Bile salts emulsify fats.
Pancreas	Pancreatic Juice	Amylase (Amylopsin)	Carbohydrates	maltose, sucrose
		Lipase	Fats	Monoglycerides, Diglycerides, Fatty acids
		Trypsin	Proteoses, Peptones	Polypeptides
Intestinal glands	Succus entericus	Erepsin (Peptidases)	Polypeptides and remaining proteins	Amino acids
		Sucrase	Sucrose	Glucose and Fructose
		Maltase	maltose	Glucose
		Lipase	Fats	Fatty acids, Glycerol

richly supplied with ***blood capillaries*** and ***lymph vessels of lacteals.***

The end products of proteins (amino acids) and carbohydrates (glucose and fructose) are absorbed into blood capillaries.

The end products of fats (fatty acids and glycerol) are absorbed into lacteals due to their large molecular size. The absorption of fats into lacteals is aided by bile salts. After absorption into the lacteals, the glycerol and fatty acids are again converted into fat globules and finally enter the blood.

ASSIMILATION

Various digested foods form the blood are taken up by tissues where they are either converted into new protoplasm or utilised for providing energy by oxidation. This phenomenon is called ***assimilation.***

EGESTION OR DEFAECATION

Undigested food along with mucous, leucocytes and dead bacteria after the absorption of digested products, reach the rectum. After the reabsorption of water in the rectum, they form the faeces which is removed from time to time through cloacal aperture. The removal of undigested waste is called ***egestion*** or ***defaecation.***

6

Circulatory System

Circulatory system is a transport system developed higher and complex animals. This system also maintains the uniformity of the internal environment of the body.

In animals like protozoans. sponges, coelenterates, flatworms, etc., the exchange and transport of materials take place by a simple process called ***diffusion***. But in higher and complex organisms, the process of diffusion becomes ineffective and inadequate due to the increase of body size. Hence, transport system is developed and it helps in the transport of various substances in the body. It is also concerned with the collection of metabolic wastes like carbon dioxide, nitrogenous wastes and conveying them to various excretory organs.

The circulatory system in frog consists of ***blood vascular system*** and ***lymphatic system***. The system of blood vessels through which the blood circulates is called ***blood vascular system***. The system of vessels through which lymph flows is known as ***lymphatic system***. The blood vascular system is closed type and that of lymphatic system is open type. The blood vascular system consists of ***heart, blood vessels*** and ***blood***. The lymphatic system comprises the ***lymph vessels*** and ***lymph***.

The blood vascular system in frog comprises ***(i) blood (ii) heart (iii) arterial system*** and ***(iv) venous system***.

BLOOD

Blood is a fluid connective tissue. It is described as ***red***

river of life. It consists of a colourless liquid matrix called, the ***plasma*** in which many ***blood cells*** or ***corpuscles*** are seen. Plasma constitutes about 55% of the blood, while blood corpuscles form the remaining 45% of the blood.

1. PLASMA

Plasma is a clear pale yellow liquid containing 90% of water and less than 10% of various organic and inorganic substances. The Inorganic substances are chlorides, sulphates, phosphates and bicarbonates of sodium, potassium, calcium, magnesium and iron. Organic substances include plasma proteins like ***fibrinogen, serum albumins*** and ***serum globulins***. Glucose, fats, amino acids, hormones, enzymes and urea are other organic substances present in plasma. Some gases like oxygen, carbon dioxide and nitrogen are found soluble in plasma. The composition of plasma generally remains constant. The pH value of plasma varies between 7.2–7.6.

2. BLOOD CORPUSCLES

The blood corpuscles are of three kinds. They are **1.** Red blood corpuscle (RBC) or erythrocytes or haematids **2.** White blood corpuscles (WBC) or leucocytes and **3.** spindle cells or thrombocytes.

Red blood corpuscles or erythrocytes

In frog, the red blood corpuscles constitute 94% of total blood corpuscles. They are oval, flattened and nucleated, but appear biconvex when seen along their edges. Each corpuscle contains cytoplasm and biconvex nucleus. An erythrocyte of frog measures about 22.3 μ in length and 15.7 μ in width. Every cubic

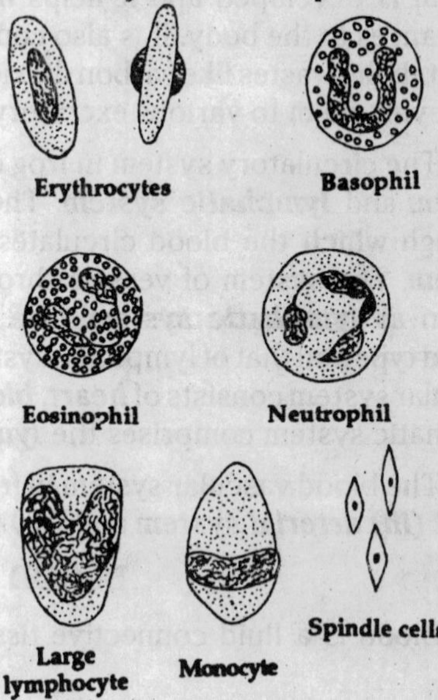

Blood Corpuscles

millimeter of blood contains about 5,00,000 red blood corpuscles. In frog, they live for about 100 days.

A ***lipoprotein matrix*** is present in ***RBC*** in the form of a network called, ***stroma***. ***Haemoglobin***, a complex respiratory pigment is found is stroma. The red colour of red blood corpuscles is due to the presence of haemoglobin. 100 ml of frog blood contains 6-10 gams of haemoglobin. It is an iron containing proteinous compound. It is the carrier of respiratory gases, the oxygen and carbon dioxide. It has great affinity with oxygen. When the haemoglobin comes in contact with oxygen in respiratory surfaces, it forms and unstable compound called ***oxy-haemoglobin***. It dissociated again into oxygen and haemoglobin in tissues of the body.

White blood corpuscles or leucocytes

White blood corpuscles are nucleated colourless cells. These are fewer in number and they constitute 0.1% of total blood corpuscles. These are irregular in shape and move with the help of pseudopodia by amoeboid movement. Their number is about 7,500 per cubic millimeter of blood. The number may increase when infection takes place in the body. This process is called ***leucocytosis***. They are described as soldiers of body as they destroy harmful micro-organisms that enter the body. So leucocytes are mainly ***protective in function***. Some of these cells are phagocytic in nature and engulf bacteria, old and dead cells or any other foreign matter. So they are called ***phagocytes***. Depending upon the shape of nucleus and the presence or absence of granules in the cytoplasm; the white blood corpuscles are divided into 2 types. ***i) Granulocytes*** and ***ii) Agranulocytes***.

i) Granulocytes or polymorphonuclear leucocytes : These cells have granular cytoplasm. The nucleus of these cells is made up of two to more lobes and these lobes are connected by a thread like structure. Granulocytes constitute 70% of the total leucocytes which are produced in spleen in frog and bone marrow in man.

Based on their staining properties, the granulocytes are divided into 3 types.

a. **Basophils** : These cells can be stained with basic dyes such as ***methylene blue*** or ***haematoxylin***. Cytoplasm of these cells has large granules and the nucleus is with two to more lobes. Basophils are non phagocytic cells and constitute 0.5% of the total leucocytes.

b. **Eosinophils or acidophils** : These cells are readily stained with acidic dyes such as eosin. The nucleus is divided into 2 lobes and the cytoplasm has many large granules. They constitute about 2%-3% of the total leucocytes. Acidophils increase in number at times of allergy and infection.

c. **Neutrophils or heterophils** : These cells are usually stained with neutral dyes. They have many granules and the nucleus is many lobed. Neutrophils are phagocytic which engulf bacteria. These cells form 67% of the total leucocytes. These are described as ***microscopic policemen.***

ii) Agranulocytes or monomorphonuclear leucocytes: The cytoplasm of these cells has no granules. The nucleus is large and undivided. These cells are produced from and undivided. These cells are produced from lymphoid tissue and constitute about 30% of the total leucocytes. The agranulocytes are of two types.

a. **Lymphocytes** : Each cell has a large round nucleus. The cytoplasm is less in quantity and non-granular. Lymphocytes are small or large in size and produced in the spleen (lymphoid tissue) and constitute about 28% of the total leucocytes. These cells defend the body by the production of antitoxins and antibodies. Lymphocytes are abundant in lymph.

b. **Monocytes** : Each cell is oval with a bean shaped nucleus. These are the largest of all leucocytes and constitute about 1%-3% of the total leucocytes. Monocytes are phagocytic cells and are produced in spleen or lymph nodes. They remove worn out and dead cells. Hence, they are described as ***internal scavangers.***

c. **Thrombocytes** : These are the spindle shaped nucleared cells. These cells are smaller then the leucocytes. The thromboplastin produced in these cells helps in blood coagulation. The thromboplastin changes the prothrom bin into thrombin. This thrombin converts the soluble fibrinogen into insoluble fibrin resulting in the clotting of blood.

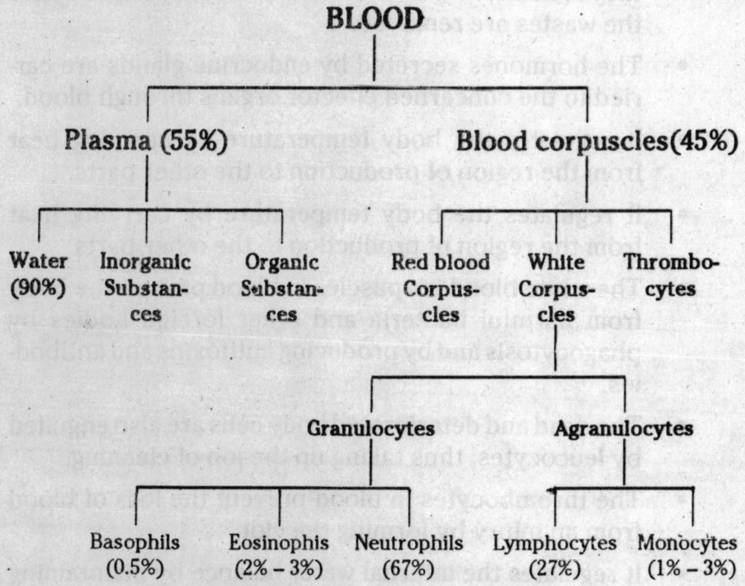

Production of Blood corpuscles

The process of formation of blood corpuscles is called ***haemopoiesis*** and the process of formation of erythrocytes alone is called ***erythropoiesis***. The tissue which produces the blood cells is the ***haemopoietic tissue***. Red blood corpuscles, white blood corpuscles and thrombocytes are produced by spleen in adult frog. During embryonic condition liver produces red blood corpuscles and both liver and spleen produce white blood corpuscels. In adult frog, these cells are produced by spleen, bone marrow and lymphoid tissue. During spring season, when there is greater metabolic activity, the bone marrow attains maximum capacity to produce blood corpuscles. Kidney is hemopoietic organ in tadpole stage.

FUNCTIONS OF BLOOD

- Blood carries all the digested food materials from the intestine to various tissues. Water, chemical substances and vitamins are also carried to the tissues.

- It transports oxygen from lungs to tissues and carbon dioxide from tissues to lungs.
- It removes metabolic wastes from various tissues and they are carried to liver and then to kidneys from where the wastes are removed.
- The hormones secreted by endocrine glands are carried to the concerned effector organs through blood.
- It regulates the body temperature by carrying heat from the region of production to the other parts.
- It regulates the body temperature by carrying heat from the region of production to the other parts.
- The white blood corpuscles in blood protect the body from harmful bacteria and other foreign bodies by phagocytosis and by producing antitoxins and antibodies.
- The dead and deteriorated body cells are also engulfed by leucocytes, thus taking up the job of cleaning.
- The thrombocytes in blood prevent the loss of blood from an injury by forming the clot.
- It regulates the internal water balance by maintaining the water content of the tissues.
- It brings all the essential substances to the site of injury for healing wounds and regeneration of body parts.

HEART

Heart is a conical, muscular, central pumping organ situated in the median line of anterior part of the body cavity. The pectoral girdle gives protection to the heart. The heart is somewhat dorsoventrally flattened. It receives blood through the veins and pumps it through the arteries to all parts of the body.

EXTERNAL STRUCTURE OF HEART

The heart of frog externally consists of ***pericardium, two auricles, one ventricle, truncus arteriosus*** and ***sinus venosus.***

Pericardium

The heart is enclosed and projected in a thin double walled bag known as ***pericardium***. It consists of an inner ***visceral layer***

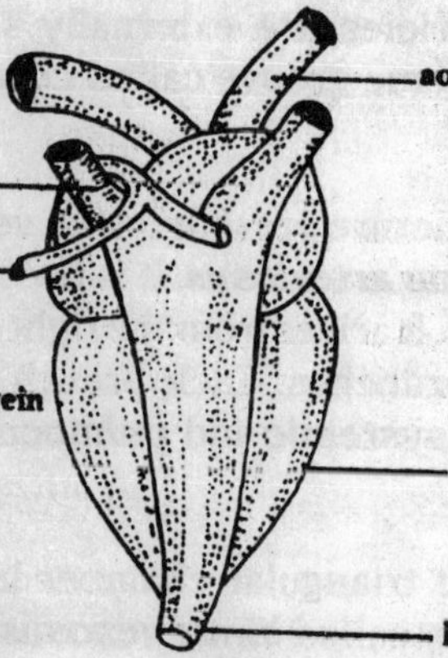

Dorsal View

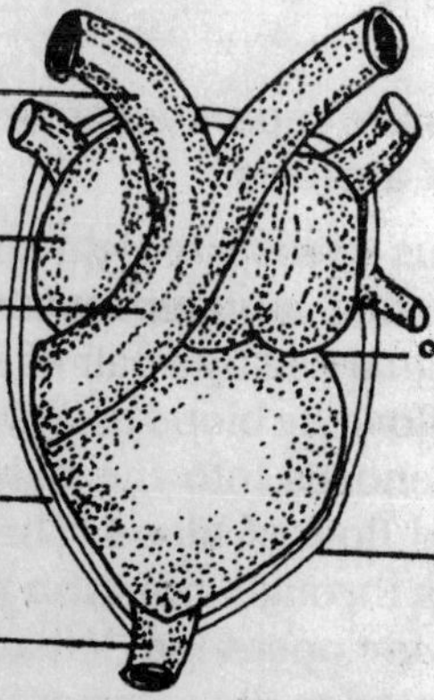

Ventral View

closely adhering the surface of heart and an outer ***parietal layer.*** Between the two layers there is a ***pericardial cavity*** filled with ***pericardial fluid.*** It prevents friction, while the heart is working. It also protect the heart from external shocks and serves as a lubricant and does not allow the heart to dry up.

Auricles and Ventricle

The heart in frog is 3 chambered consisting of ***2 auricles (left and right auricles)*** and single ***ventricle.*** The right auricle is larger than the left. The ventricle is a thick walled conical structure. The 2 auricles are externally separated from the ventricle by a transverse groove called ***coronary sulcus.***

Truncus Arteriosus

The tubular structure present on the ventral surface of the heart is called ***truncus arteriosus.*** It is the largest blood vessel of the arterial system. It arises from the right side of the ventricle and divides into 2 branches. Each branch gives rise to three trunks viz., ***carotid, systemic*** and ***pulmocutaneous.***

Sinus Venosus

The thin walled triangular chamber laying on the dorsal surface of the heart is called ***sinus venosus.*** It is formed by the union of two ***precaval veins*** and one ***post caval vein.***

INTERNAL STRUCTURE OF HEART

Auricles

The two auricles are separated by an ***inter-auricular septum*** into right and left auricles.

The sinus venosus opens into right auricle through a transverse oval opening called ***sinu-auricular aperture.*** The sinu-auricular aperture is guarded by a pair of ***sinu-auricular valves*** whose function is to allow the blood to flow only in one direction. i.e., from the sinus venosus into the right auricle. Thus, they prevent the backward flow of blood. The left auricle receives pure blood from lungs through ***common pulmonary vein.*** This common pulmonary vein opens into left auricle through a small opening which is situated in the anterior wall of the left auricle. The opening is not guarded by any value.

Ventricle

The anterior thin walled auricles are separated from posterior thick walled ventricle by a partition called ***auriculo-ventricular septum.*** The two auricles open into ventricle by an ***auriculo-ventricular aperture.*** This aperture is bounded by two pairs of membranous flaps called ***auriculoventricular valves,*** one arising from the dorsal edge and the other from ventral edge of this aperture. These valves are connected with the muscular wall of the ventricle by fine threads of collagen fibres called the

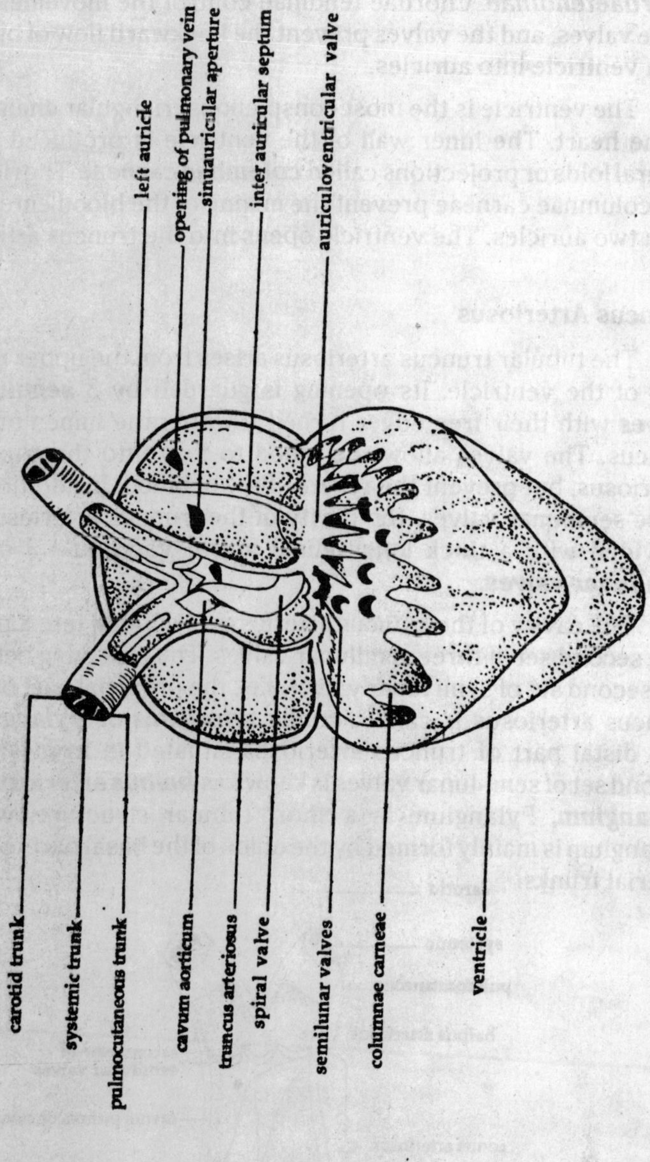

Internal Structure of Heart

Chordaetendinae. Chordae tendinae control the movement of these valves, and the valves prevent the backward flow of blood from ventricle into auricles.

The ventricle is the most conspicuous triangular chamber of the heart. The inner wall of the ventricle is produced into several folds or projections called ***columnae carneae***. The ridge-like columnae carneae prevent the mixing of the blood entering from two auricles. The ventricle opens into the truncus arteriosus.

Truncus Arteriosus

The tubular truncus arteriosus arises from the upper right side of the ventricle. Its opening is guarded by 3 ***semilunar valves*** with their free edges turned towards the lumen of the truncus. The valves allow the blood to flow into the truncus arteriosus, but prevent the reverse flow of blood. In addition to these semilunar valves, the cavity of the truncus arteriosus is provided with a thick longitudinal ***spiral valve*** and 3 more ***semilunar valves.***

The cavity of the truncus arteriosus is divided into 2 parts by a second set of three semilunar valves. The part lying behind the second set of semi lunar valves, i.e., the proximal part of the truncus arteriosus is called, ***conus arteriosus*** or ***pylangium***. The distal part of truncus arteriosus situated in front of the second set of semi-lunar valves is known as ***bulbus arteriosus or synangium,*** Pylangium is a short tubular structure, while synangium is mainly formed by the union of the basal parts of the arterial trunks.

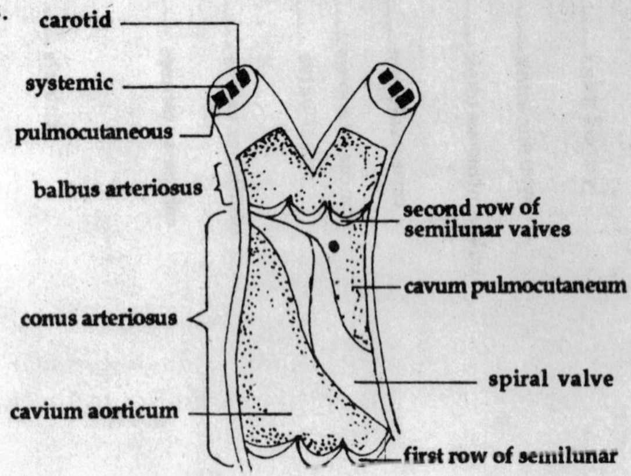

Truncus Arteriosus

The **spiral valve** is a ribbon like spirally twisted longitudinal fold which hangs freely from the dorsal wall of the conus arteriosus. It incompletely divides the cavity of conus arteriosus into two parts, a dorsal and left ***cavum pulmocutaneum (pulmocutaneous cavity)***, and ventral and right ***cavum aorticum (systemo-carotid cavity)***. The spiral valve helps in determining the course of blood leaving the heart. It also gives mechanical support to the truncus arteriosus.

The truncus arteriosus anteriorly divides into a left and right branches. Each branch gives rise to 3 aortic arches or trunks, viz., ***1. the anterior carotid arch, 2. the median systemic arch, and 3. posterior pulmo-cutaneous arch***.

FUNCTIONING OF HEART

The muscular heart contracts and expands at regular intervals. The contraction of the heart is called systole while its expansion is known as diastole. One ***systole*** and one ***diastole*** constitute a heart beat and the beating occurs rhythmically. The animal dies, if the beating of heart is stopped.

The contraction of heart starts from the ***sinu-auricular node***. It is a patch of muscles situated between sinus venosus and right auricle. It initiates the contraction and maintains the pace. The specialised part from where the contraction begins is called ***myogenic pace maker***. Such a heart with myogenic pacemaker is called ***myogenic heart***. The electric oscillations produced from pace maker spread to the muscles of the heart and initiate contraction.

Several theories have been put forward by scientists to explain the course of circulation and working of the heart. There are two main theories explaining the course of blood circulation in the heart of frog:

A. ***Classical theory*** and B. ***Modern theory***.

Classical Theory or Old Hypothesis

This classical theory was proposed by **Brucke** in 1858 and later it was supported by **Sabatier** in 1873. This theory was held valid upto 1950 i.e.., nearly hundred years.

According to this theory, the heart of frog shows ***incomplete double circulation***, with a complicated mechanism of separating the pure and impure blood.

According to **Brucke**, a the sinus venosus first contracts, so that the impure blood enters right auricle through the sinu-auricular aperture. At the same time, the left auricle receives the pure blood from lungs through pulmonary veins. Both auricles contract almost simultaneously sending the blood into the ventricle through the auriculoventricular aperture. Thus the left side of the ventricle receives pure blood from the left auricle while the right side of the ventricle receives impure blood from the right auricle. However, the mixing of pure and impure blood is prevented because of the ***viscous nature of blood and the spongy nature of ventricular wall.*** Moreover the ventricle contracts quickly without allowing the blood to mix. Yet, there is mixing of the two types of blood in the middle region of the ventricle.

When the ventricle contracts the impure blood first enters cavum pulmocutaneum of truncus arteriosus and finally reach lungs and skin through pulmocutaneous arch. Later the mixed blood enters the cavum aorticum and then into systemic arches which supply the blood to limbs and visceral organs. Finally the pure or oxygenated blood flows into the carotid arches through the cavum aorticum. The carotid arches supply pure blood to brain and some other parts of the head. According to this theory the spiral valve in the conus arteriosus is believed to regulate the flow of blood into 3 aortic arches on each side. It was also thought that the blood pressure in the three aortic arches was different.

The impure blood flows first into pulmocutaneous arch as it offers least resistance. Moreover, it is the shortest route which is devoid of any obstructions. The mixed blood takes its course into systemic arches because they offer less resistance due to their wide lumen. Moreover the carotid glands (carotid labyrinths) resist the flow of impure and mixed blood into the carotid arches. Finally, the pure blood enters carotid arches as there is no other alternative.

This explanation of flow of blood into aortic arches from the ventricle is now discarded due to lack of experimental evidence.

Modern theory

Modern theory was propounded by **Vandervael** and **Foxon**. The recent experiments conducted by them clearly show that the blood from the lungs and various other parts of the body is mixed.

This mixed blood that is collected into ventricle flows into all the three aortic arches simultaneously but not one after the other. In 1933 **Vandervael** destructed the connection between the brain and spinal cord (pitching) and injected **Indian ink particles** into the pulmonary veins. Then he studied the flow of blood by illuminating the heart and blood vessels from below. He found the flow of Indian ink simultaneously through the three aortic arches. The observations made by **Vandervael** were confirmed by **Foxon**. In 1958, Foxon injected a radio-active opaque sub-stance called ***thorostrat*** (a colloidal suspension of the thorium dioxide) into the pulmonary veins.

The radiographs taken with an interval of **15-30** seconds by **Foxon**, proved that the blood flows simultaneously into all the three aortic arches. Thus there is no selective flow of oxygenated blood to the head region. When the blood pressure in the three aortic arches of frog was measured by ***optical electronic manometer***, it was found that there is no difference in the blood pressure. In contrast to the findings of Foxon and Vandervael, it was revealed by **Parker** and **Haswell** (1963) that blood flows into the three aortic arches by differential distribution but it is not still known how it takes place.

In the light of the recent experiments we can conclude that—

- The lungs in frog are not important respiratory organs. Respiration mainly takes place by the skin even when the frog is on land. Moreover the pulmonary respiration does not occur when frog is in water. So the oxygenation of blood is mainly brought about by skin.
- The left auricle receives only mixed blood as lungs are not so efficient respiratory organs.
- The blood which comes into sinus venosus from the skin and bucco pharyngeal cavity is oxygenated while the blood that comes from other parts of the body is deoxygenated. So, the blood that enters the right auricle from the sinus venosus is mixed in nature.
- The mixed blood from both the auricles gets mixed thoroughly in the ventricle. The longitudinal ridges or columnae carneae of the ventricle do not prevent the

mixing of blood. They increase the inner surface area enabling it to receive more blood and oxygen.

- The spiral valve has no regulatory function. It provides only mechanical support to the truncus arteriosus.
- The spiral valve has no regulatory function. It provides only mechanical support to the truncus arteriosus.
- The carotid labyrinth present at the base of each carotid artery has a sensory functions i.e.., detects the pressure, chemical changes and oxygen tension of blood.
- The flow of the blood into all the three aortic arches is simultaneous.

The heart receives both oxygenated and deoxygenated bloods which get mixed inside the heart. So, the heart pump mixed blood through the arteries. This type of circulation is intermediate between the single and double circuits.

Blood Vessels

The tubes that carry blood are blood vessels. These are of two types.

1. ***Arteries*** and 2. ***Veins***.

Arteries

The blood vessels which carry oxygenated blood (except pulmonary artery) away from the heart to different parts of the body are called ***arteries***. They have thick and more elastic walls. Each artery is made up of three layers.

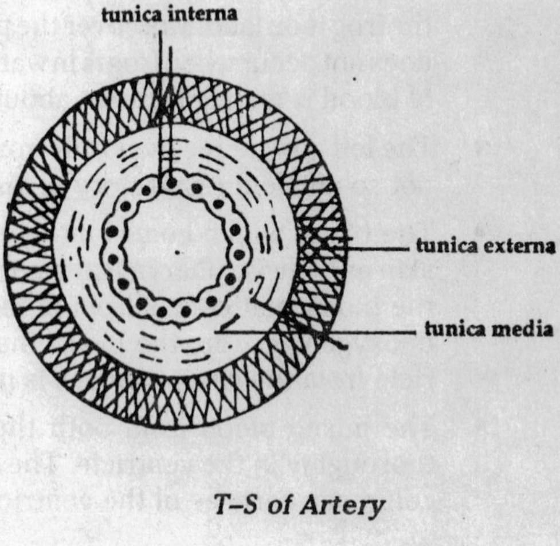

T–S of Artery

i) Tunica

externa or adventia : It is the outermost layer which is made up of loosely arranged elastic and collagen fibres.

ii) **Tunica media** : It is the thick middle layer which is made up of smooth muscle fibres and elastic connective tissue fibres. This layer is greatly developed in arteries.

iii) **Tunica interna or intima** : It is the inner most layer which is made up of flat endothelial cells. Internally it is lined by an elastic membrane.

The blood flows at a relatively greater speed in arteries. The lumen of these blood vessels is narrow and devoid of valves. The arteries are relatively deep seated in the body. They get divided into fine branches called ***arterioles*** which break up into capillaries. Thus arteries end with capillaries.

Veins

The blood vessels which carry blood from different parts of the body (except pulmonary veins) towards heart are called veins.

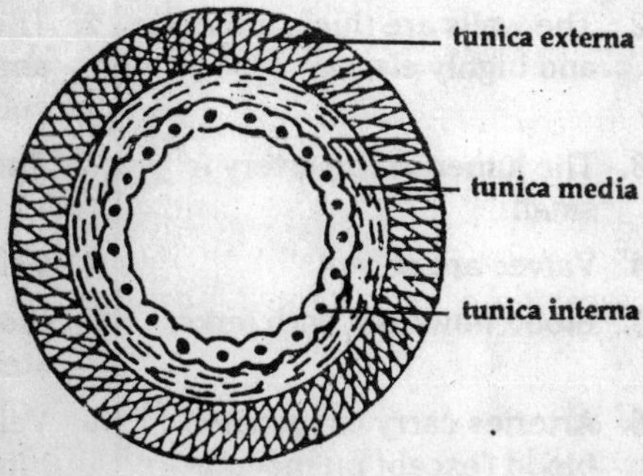

T–S of Vein

These vessels also composed of the same three layers as in arteries. The ***tunica externa*** occupies the greater part of the wall of a vein. The collagen fibres of the outer layer or tunica externa are more developed. The muscles and elastic connective tissues are less developed in ***tunica media***. The ***tunica interna*** lacks the elastic membrane. The walls of the veins are much thinner than the arteries. The lumen of the veins is large and contains valves. Veins are relatively superficial in the body. The blood moves slowly in the veins and backward flow of blood is prevented by valves. Capillaries unite to form ***veinules*** and veins are formed by the union of several veinules.

i) **Tunica externa or adventia** : It is the outermost layer which is made up of loosely arranged elastic and collagen fibres.

ii) **Tunica media** : It is the thick middle layer which is made up of smooth muscle fibres and elastic connective tissue fibres. It is made up of a single layer of flattened epithelial cells. The arteries are connected with veins through capillaries. The capillaries are generally found in tissues in the form of a net work.

COMPARISON OF ARTERIES AND VEINS

Arteries	*Veins*
1. Arteries carry blood away from the heart.	1. Veins carry blood towards the heart.
2. The walls are thick, muscular and highly elastic	2. They have thin walls. Muscles and elastic fibres are less developed
3. The lumen of the artery is small.	3. The lumen of the vein is large.
4. Valves are absent.	4. Valves are present.
5. Blood flows fast with jerks.	5. Blood flows slowly and steadily.
6. Arteries carry oxygenated blood (except pulmocutaneous artery) from the heart to different parts of the body.	6. Veins carry deoxygenated blood (except pulmonary vein and cutoaneous vein) to the heart from different parts of the body.
7. They are deep seated blood vessels in the body.	7. They are superficial blood vessels in the body.
8. The blood is bright scarlet red coloured	8. The blood is bluish-red coloured.
9. They remain unchanged even after the loss of blood.	9. Veins collapse after the loss of blood.
10. Inner layer is provided with an elastic layer.	10. Inner layer lacks elastic membrane.

ARTERIAL SYSTEM

All the blood vessels which carry the blood away from the heart to different parts of the body are called ***arteries*** and they constitute the ***arterial system***. The arterial system in frog begins with ***truncus arteriosus***. It is divided into large right and left branches called ***aortic trunks***. Each trunk gets divided into three aortic arches. They are 1. The anterior ***common carotid***. 2. The middle ***systemic***. 3. The posterior ***pulmo-cutaneous***.

CAROTID ARCH

The carotid arch of each side divides into two, and ***external carotid*** or ***lingua artery*** supplying blood to the tongue and the lower jaw and an ***internal carotid artery*** to the orbit and brain. Each internal carotid at its base bears a swollen ***carotid labyrinth*** or ***carotid gland***. It has a network of capillaries and ***chromoffin cells***. Carotid gland or labyrinth detects the pressure and chemical changes in blood and regulates the blood pressure. So it is called ***baro and chemoreceptor***. The internal carotid runs forward and gives rise to a ***palatine artery*** supplying blood to the roof of the bucco pharyngeal cavity. The carotid arch on each side gives off a small artery from its base called ***coronary artery*** or ***arteria bulbi*** supplying blood to the heart muscles.

SYSTEMIC ARCH AND ITS BRANCHES

The systemic is the largest middle branch of the truncus arteriosus. The right and the left systemic arches curve around the oesophagus, proceed backwards and join dorsally to form median vessel known as ***dorsal aorta*** which supplies blood to different parts of the body. Before the formation of dorsal aorta, each systemic gives off three branches *viz.*, ***occipito-vertebral artery, sub clavian artery and oesophageal artery***.

a. **Occipito-vertebral artery** divides into an occipital artery supplying the blood to posterior part of the skull and ***vertebral artery*** supplying blood to the muscles of the vertebral column.

b. **Sub-calvian artery** carrying blood to the fore limbs, and

c. **Oesphageal artery** supplying the blood to the oesophagus.

The right systemic continues behind as the ***dorsal aorta.*** The left systemic continues as ***coeliaco-mesenteric artery*** carrying blood to alimentary canal and digestive glands like liver and pancreas. The two systemics are communicated with each other through a small pore.

The coeliaco-mesenteric artery passing through the mesentery divides into two main branches, a small ***coeliac artery*** supplying blood to the stomach, pancreas, liver (gastric artery) and a ***large mesenteric artery***. The large mesenteric artery divides into and ***anterior mesenteric*** and a ***posterior mesenteric artery***. The anterior mesenteric carries blood to the duodenum ***(duodenal artery)***, spleen ***(spleenic artery)*** and ileum ***(intestinal artery)***. The posterior mesenteric artery supplies blood to the large intestine or rectum.

The ***dorsal aorta*** runs back wards between the kidneys along the mid-dorsal line. It gives off four to six pairs of ***renal arteries*** supplying blood to the kidneys. The first pair of renal arteries gives off gonadial arteries (ovarian in female and ***spermatic*** in male) to the gonads.

Posteriorly the dorsal aorta divides into two large ***iliac arteries*** running towards the hind limbs. Each iliac artery gives a ***vesciculoepigastric artery*** supplying blood to the urinary bladder and wall of abdominal cavity. A ***femoral*** or ***external iliac*** arises from the iliac artery which supplies blood to the outer and dorsal part of the thigh. The iliac artery continues as the ***sciatic or internal iliac*** carrying blood to the inner or ventral part of the hind limb.

PULMO-CUTANEOUS ARCH

This is the posterior branch of the aortic trunk. The pulmocutaneous arch divides into a ***pulmonary artery*** which carries blood to lungs and a ***cutaneous artery*** carrying blood to the skin for oxygenation.

VENOUS SYSTEM

All the blood vessels which collect the blood from different parts of the body and carry the same to the heart are called ***veins***. The veins constitute the venous system. The origin for the formation of veins is capillaries.

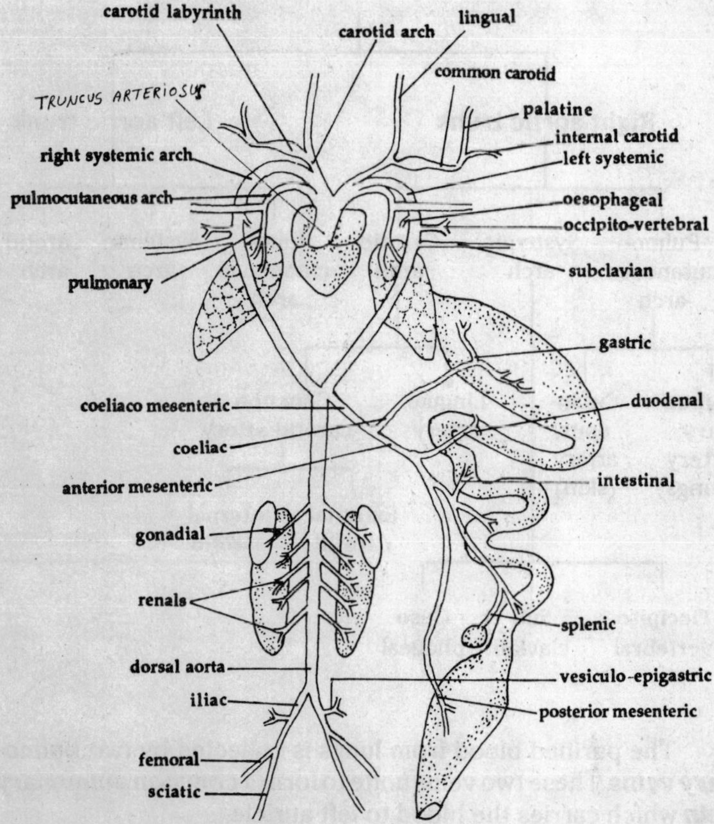

Arterial system

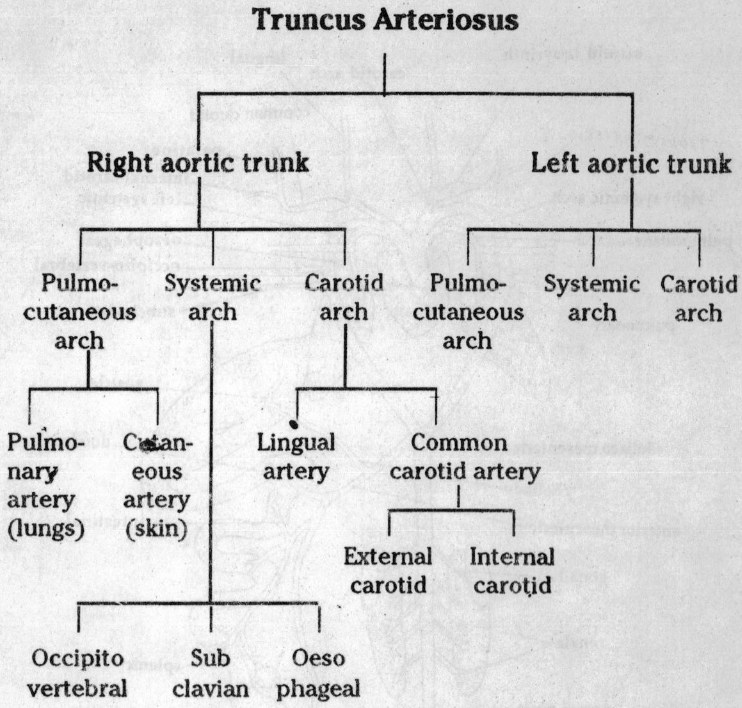

The purified blood from lungs is collected by two ***pulmonary veins***. These two veins unite to form a ***common pulmonary vein*** which carries the blood to left auricle.

The blood from the rest of the body comes to ***sinus-venosus*** which opens into right auricle through sinu-auricular aperture. The sinus venosus is formed by the union of two ***pre-caval veins*** and one ***post.caval vein***. The blood from the anterior part of the body is collected by two ***pre-cavál veins*** or ***anterior vena cavae*** and from the posterior parts is collected by a ***postcaval vein*** or ***posterior vena cava***. These three major veins open into sinus venosus.

PRECAVAL VEIN

It is formed of three veins namely, 1. ***External jugular vein***, 2. ***Innominate vein*** and 3. ***sub-calvian vein***.

External jugular vein is formed by the union of (a) ***lingual vein*** bringing blood from the tongue and (b) ***mandibular vein*** bringing blood from the lower jaw.

The innominate vein is formed by the union of (a) ***internal jugular vein*** bringing blood from the brain the orbit and (b) ***subscapular vein*** bringing blood from the muscles of the scapula (shoulder).

The left innominate receives impure blood from the muscles of the heart through a branch called ***vena bulbi anterior.***

Subclavian vein : It is formed by the union of (a) ***branchial vein*** collecting blood formed the fore limb, and (b) ***musculocutaneous vein*** collecting blood from the skin the muscles of the abdomen and also mucous membrane of mouth and head muscles.

The arrangement of various veins in the anterior region is given below:

1. **External jugular**	a. **Lingual** (tongue) b. **Mandibular** (lower jaw)
2. **Innominate**	a. **Internal jugular** (brain and orbit) b. **Subscapular** (shoulder)
3. **Subclavian**	a. **Musculo-cutaneous** (muscles of the body wall and skin) b. **Branchial** (Fore limb).

POSTCAVAL VEIN

The postcaval vein arises between the two kidneys and it is formed by the union of four pairs of ***renal veins and the gonadial veins***. The renal veins collect blood from kidneys. The first pair of renal veins receive a ***gonadial vein*** (***spermatic*** in male and ***ovarian*** in female) each.

The postcaval vein runs forwards, passes through the liver and opens into the posterior end of sinus venosus. On its way, it receives two ***hepatic veins*** which collect blood from the liver.

In addition to the veins mentioned above, there are two

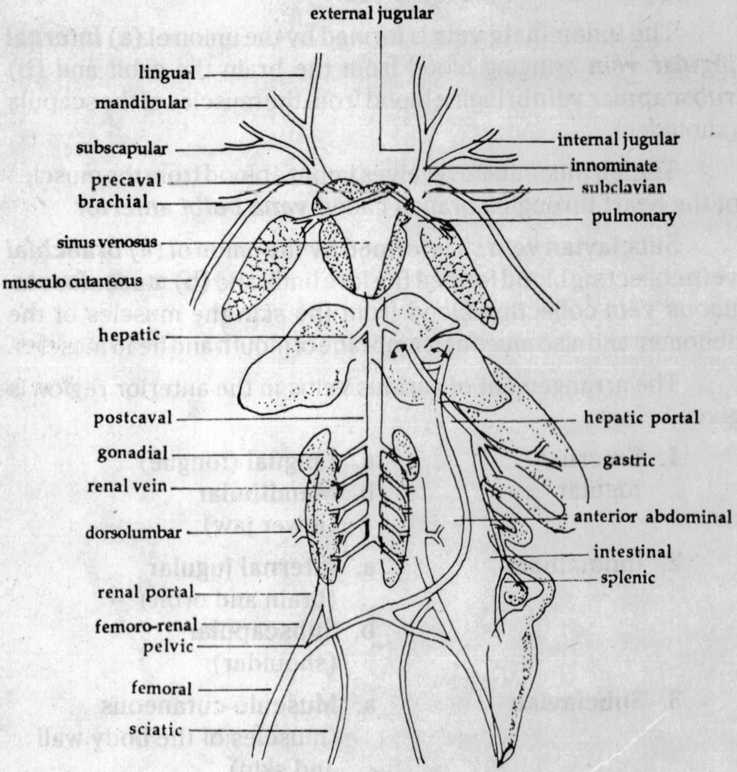

Venous System

special venous paths, the ***renal portal system*** and the ***hepatic portal system.***

PORTAL SYSTEM

Portal system is formed by portal veins. The vein which starts with capillaries in an organ and ends with capillaries in another organ other than heart is called ***portal vein***. The portal veins are both ***collecting and distributing vessels***. Both renal and hepatic portal systems are well deveioped in frog.

RENAL PORTAL SYSTEM

The veins which collect the blood from the posterior end of the body (hind limbs) constitutes the ***renal portal system***. The blood form each leg is collected by two large veins, a) the ***femoral vein*** from the outerside of the thigh and b) ***sciatic vein*** from the inner side of the thigh. The femoral vein, as it leaves the thigh divides into ***femero-renal*** and ***pelvic veins***. The femero-renal vein runs forwards and unites with the sciatic vein to form ***renal portal vein***. It enters the kidney of its side and break up into capillaries. It is to be noted that the ***renal portal system starts in capillaries in the kidneys***. Before entering the kidney the

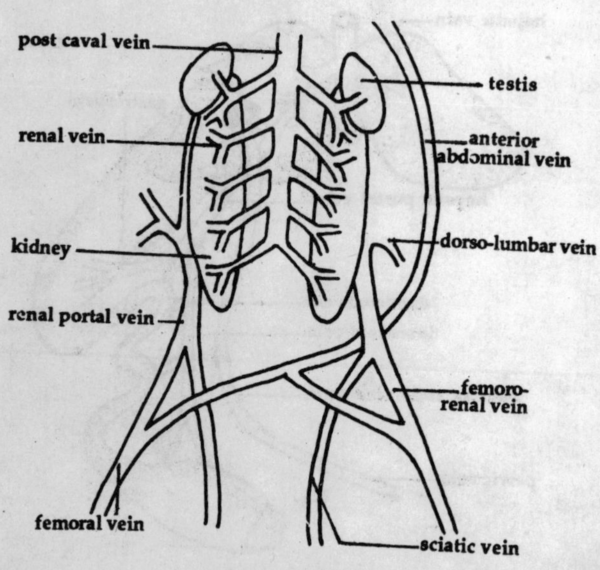

Renal Portal System

renal portal vein receives the ***dorsolumbar vein*** collecting blood from the dorsal body wall. Thus, the blood from hind limbs is not directly collected by the post caval vein but it is collected by the renal portal veins which empty the blood into kidneys. The system separates the nitrogenous waste from blood and supply the same to kidney for removal.

HEPATIC PORTAL SYSTEM

The hepatic portal system includes two veins namely a) ***the anterior abdominal vein*** and b) ***the hepatic portal vein.*** The anterior abdominal vein is formed posteriorly in the abdomen by the union of ***the right and left pelvic veins.*** It runs forwards along the midventral line of the abdomen and reaches liver. The anterior abdominal vein is included in the hepatic portal system because it reaches the liver and breaks up into capillaries. Before it enters the liver, it receives, ***i) a vesicular vein*** collecting blood from the urinary bladder, ***ii) a parietal vein*** bringing blood from the ventral body wall and ***iii) a cardiac vein*** or ***vena bulbi posterior*** collecting blood from the muscles of the heart.

The anterior abdominal vein before entering liver gives off a branch which is connected with hepatic portal vein.

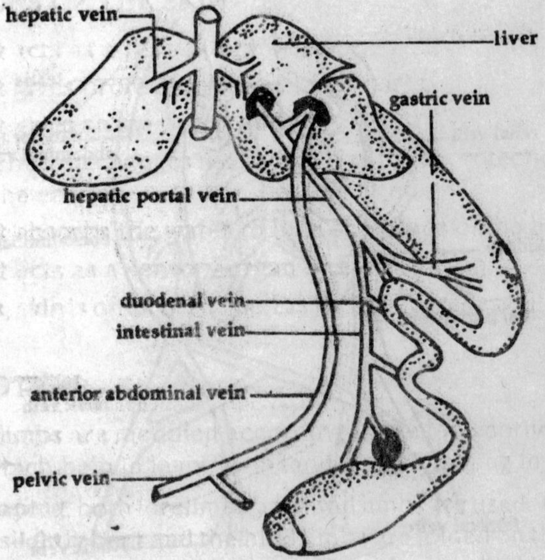

Hepatic Portal System

The hepatic portal vein is formed by the union of several veins bringing blood from different parts of the alimentary canal, i.e., from stomach (***gastric vein***), intestine (***intestinal vein***) pancreas (***pancreatic vein***) and spleen (***spleenic vein***). Finally the hepatic portal vein breaks up into capillaries in liver.

Thus the blood from the digestive system is collected by the hepatic portal vein and empties the same into liver. This system separates the excess food materials (glucose) from the blood and store them in liver (glycogen).

The Differences between Hepatic and Renal Portal Systems

Hepatic Portal System	***Renal Portal System***
1. The veins which collect the blood from different parts of alimentary canal and carry blood to liver constitute hepatic portal system.	1. The veins which collect the blood from posterior side of the body and carry blood to constitute the renal portal system.
2. Hepatic portal vein is formed by the union of several veins bringing blood from different parts of the alimentary canal i.e. from stomach (gastric vein), intestine (intestinal vein), pancreas (pancreatic vein) and spleen (spleenic vein).	2. Renal portal vein is formed by the union of femor-renal vein (branch of femoral vein) and sciatic vein (form inner side of thigh).
3. Hepatic portal vein terminate with capillaries in the liver.	3. Renal portal vein terminates with capillaries in the kidney.
4. Hepatic portal vein communicates with anterior abdominal vein before it terminates with capillaries in the liver.	4. Renal portal vein receives a dorso-lumbar vein from the dorsal body wall before it enters the kidney.
5. The blood from alimentary canal is carried to the liver by hepatic portal vein where carbohydrates are absorbed. Excess carbohydrates are changed into glycogen in liver.	5. The nitrogenous wastes present in blood are removed when it is carried to kidney by renal portal vein from hind limbs.

LYMPHATIC SYSTEM

The lymphatic system is closely associated with blood vascular system. It consists of ***lymph, lymph capillaries, lymphatic vessels, lymph sinuses and lymph hearts***. True lymphatic vessels comparable to blood vessels are absent in frog.

LYMPH

The blood is carried to different parts of the body by the arteries. When the blood flows through capillaries, certain portions of the fluid plasma along with leucocytes of the blood passout through them as they have very thin walls. After the exchange, most of the fluid is absorbed by blood capillaries and some of the fluid remains between the tissue cells. This squeezed fluid from the blood is known ***as lymph. In short, the lymph is the blood without red blood corpuscles and some plasma proteins.***

Blood – R.B.C + Blood proteins = Lymph

Lymph is the medium of exchange of materials between the blood and tissue cells. It is considered as the ***middle man*** between blood and tissues. It fills the body spaces and bathes the tissue cells where it is known as ***tissue fluid***. It carries food and oxygen to the cells and receives waste substances from them.

LYMPH CAPILLARIES

Extending between the intercellular spaces of tissues and blood capillaries, there are thin tubes blindly ending in a mesh. These are called ***lymph capillaries***. The swollen knobs of these capillaries help in draining the lymph from the tissues.

LYMPH VESSELS

Lymph capillaries unite to form lymph vessels. These vessels bear ***lymph nodes*** at certain regions. These lymph nodes produce ***lymphocytes***. The lymph vessels of intestine are called ***lacteals***. They help in absorption of fatty acids and glycerol. The lymph vessels open into ***lymph sinuses*** or ***spaces***.

LYMPH SINUSES

There are about twenty ***lymph sinuses*** below the skin. They are called ***subcutaneous lymph sinuses***. The skin in frog

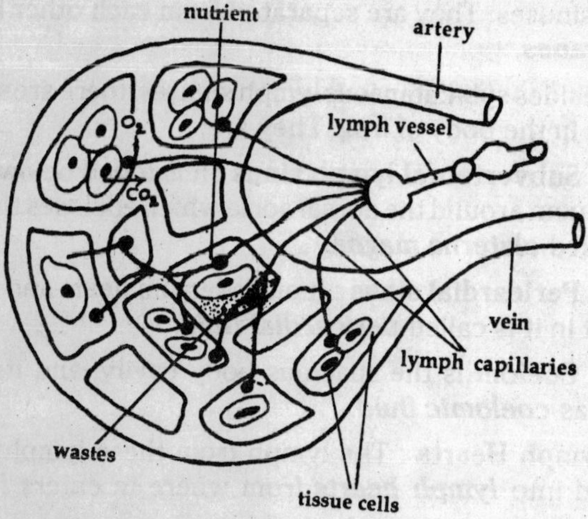

Bed of lymph capillaries

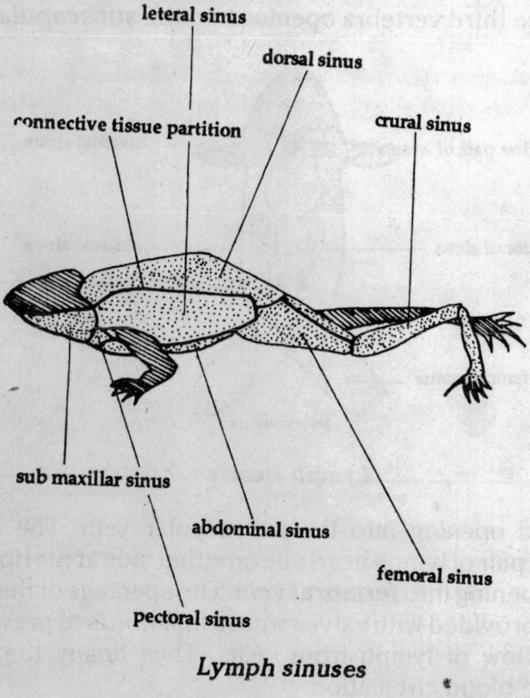

Lymph sinuses

is loosely attached to muscular body wall due to the presence of lymph sinuses. They are separated from each other by ***fibrous membranes.***

Besides sub-cutaneous lymph sinuses, there are some other sinuses in the body of frog. They are;

1. **Subvertebral lymph sinus** : It is found below the vertebral column around the dorsal aorta which encloses the kidneys. It is called ***cisterna magna.***

2. **Pericardial sinus** : It encloses the heart and the lymph present in it is called ***pericardial fluid.***

3. **Coelom** is the spacious body cavity, and its lymph is known as ***coelomic fluid.***

Lymph Hearts : The lymph from these lymph sinuses is pumped into ***lymph hearts*** from where in enters finally into veins.

There are two pairs of ***lymph hearts*** in the body of frog. The first (anterior) pair is situated just behind the transverse processes of the third vertebra opening into the ***subscapular vein.***

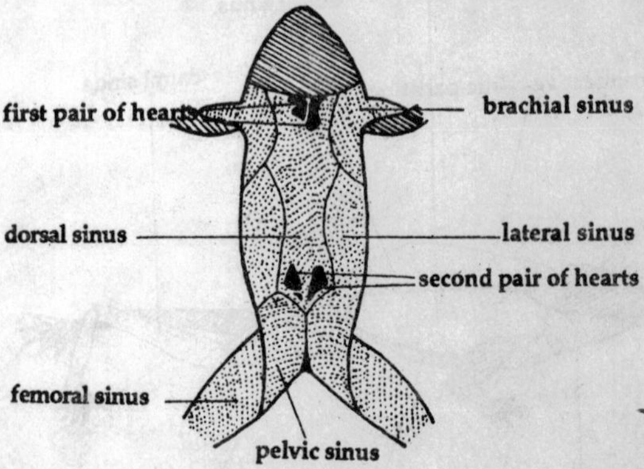

Lymph Hearts

The second opening into the subscapular vein. The second (posterior) pair of lymph hearts lie on either side at the tips of the urostyle, opening into ***fermoral vein.*** The openings of the lymph hearts are provided with valves whose function is to prevent the backward flow of lymph from veins. Thus finally the lymph reaches the blood circulation.

The Differences between Blood and Lymph	
Blood	***Lymph***
1. Blood consists of plasma and three types of blood corpuscles namely R.B.C., W.B.C. and thrombocytes.	1. The lymph consists of blood without R.B.C. and blood proteins.
2. It is the part of blood vascular system.	2. The flow of lymph constitutes lymphatic system.
3. The blood flows in heart, arteries, veins and capillaries.	3. The lymph flows in lymph vessels, lymph sinuses and lymph hearts.
4. The blood is the transporting medium for various substances in the body.	4. Lymph is the middle man between blood and tissues.
5. Blood constitutes the main drainage system in the body.	5. Lymph constitutes suplementary drainage system in the body.
6. Blood could not pass out through the capillaries into the tissues.	6. The lymph fill the body spaces and bathes the tissue cells.

SPLEEN

It is a globular dark red coloured structure which is situated at the anterior end of the large intestine. It is made up of connective tissue and muscle fibres covered with *fibrous capsule*. The capsule is produced inwards to form *trabeculae* which divide the spleen into lobules.

Functions of Spleen

1. It acts as reservoir of blood pumping the blood into circulation whenever needed.
2. It act as an ***haemopoietic organ*** as it produces red blood cells and white blood cells.
3. It destroys the worm out RBC and thus acts as ***grave yard of erythrocytes*** or ***haemolytic organ***. It is the largest lymphatic organ.

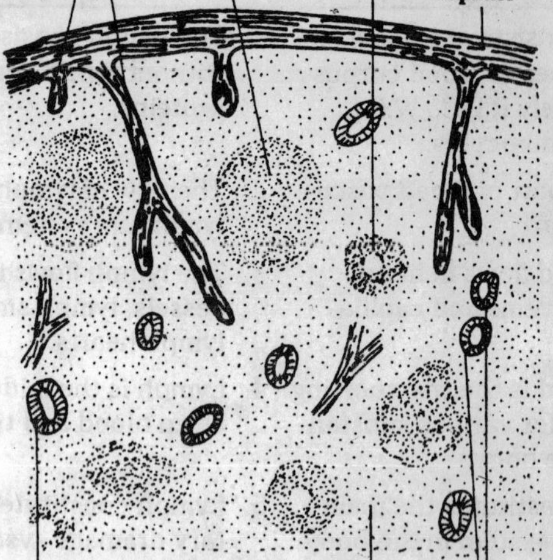
white pulp
trabeculae with blood vessels
red pulp
fibrous capsule
artery
splenic pulp
splenic nodules

T–S. of Spleen

7

Respiratory System

The catabolic process during which the oxidation of the food substances takes place in the body to release energy is called ***respiration***. It is represented as a biochemical reaction given below.

$$C_6H_{12}O_6 + 6\,O_2 \rightarrow 6Co_2 + 6H_2O + \text{energy}$$

This process includes the exchanges of gases, namely Oxygen and carbon dioxide. The organs required for this exchange of gases constitute the respiratory system. Respiration is of two types, ***aerobic*** and ***anaerobic***. Oxygen is required in aerobic respiration and it is not required in anaerobic respiration.

Respiration is performed in three phases *viz*., ***breathing (external respiration), circulatory phase and oxidation (internal respiration).*** Breathing is a mechanical process of taking atmospheric air into the respiratory organs and sending out the carbon dioxide from the body. It is also called ***external respiration or ventilation.***

Circulatory phase involves the transport of Oxygen to the tissues by respiratory pigments which are found in blood. During the ***oxidation*** (cellular oxidation) the food substances are broken down with the involvement of Oxygen, and energy is released along with the formation of water and carbon dioxide. The energy is stored in cells in the form of adenosine triphosphate (ATP) for ready use.

Frog is an amphibious animal and possesses well developed respiratory organs. Respiration is carried out by three methods. They are : 1) Cutaneous respiration 2) Bucco-pharyngeal respiration and 3) Pulmonary respiration.

CUTANEOUS RESPIRATION

- The skin is an important respiratory organ of frog. The respiration carried out by the skin is known as ***cutaneous respiration***.
- The skin functions as one of the respiratory organs during normal conditions. But the skin is the only respiratory organ during hibernation and aestivation.
- The skin of frog is always kept moist and slimy due to the presence of mucous secreting glands. The skin is richly supplied with blood capillaries of cutaneous artery.
- On land, the atmospheric Oxygen mixes first with the mucous layer. The Oxygen diffuses later into the blood capillaries. The blood carries Oxygen to various tissues in the form of oxyhaemoglobin.
- The carbondioxide formed in the tissues is carried into blood capillaries of skin from where it diffuses out.
- The Oxygen uptake by the skin of frog is about 50 ml/kg/hr throughout the year and elimination of Carbon dioxide by skin exceeds that of the lungs all round the year.
- The cutaneous respiration accounts for 35% of the respiration in frog. ***The frog dies if the skin dries up.***

BUCCO-PHARYNGEAL RESPIRATION

- This is carried out by the mucous membrane lining the bucco-pharyngeal cavity. The bucco-pharyngeal cavity is kept moist and is richly supplied with blood vessels.
- The bucco-pharyngeal respiration is carried out by lowering and raising the floor of the bucco-pharyngeal cavity. The frog at rest lowers and raises the buccal floor rhythmically about 80-120 times per minute.
- The mouth is kept closed during this process. When the floor of the buccopharyngeal cavity is lowered air

enters the buccopharyngeal cavity through nostrils.

- The oxygen of the air dissolves in mucous and diffuses into blood capillaries. Simultaneously, carbondioxide diffuses into bucco-pharyngeal cavity and is sent out through nostrils when the buccal floor is raised.
- The bucco-pharyngeal respiration is not of much use to frog. This is due to the fact that the bucco-pharyngeal cavity forms only 0.9% of that total capillary surface area of the respiratory organs put together. So, this respiration accounts to only 1% of total respiration in frog. When the mouth of frog is kept open for a long time, it will die due to lack of respiration.

PULMONARY RESPIRATION

The respiration performed by lungs is referred to as ***pulmonary respiration***. It takes place when the frog is outside the water.

Organs of Pulmonary Respiration

- The organs of pulmonary respiration are external nostrils, internal nostrils, laryngotracheal chamber and lungs.
- A pair of ***external nostrils*** is present at the tip of the snout which open into nasal cavities. The nasal cavities open into bucco-pharyngeal cavity through the ***internal nostrils.***
- The bucco-pharyngeal cavity opens through ***glottis*** into a short wide chamber called ***laryngo-tracheal chamber***. This chamber is supported by five cartilages which includes ***circular cricoid, a pair of semicircular arytenoids and a pair of pre-arytenoids.***
- The **cricoid** is a ring like cartilage surrounding the laryngo-tracheal chamber. The cricoid forms a median spine and two backwardly directed lateral ***tracheal processes***. These two processes join to form a loop like structure. The lungs are attached to the tracheal processes of cricoid. The cricoid is also attached to the posterior horns of hyoid apparatus by ligaments.

The **arytenoids** are semi-lunar cartilages lying on either side of glottis. The muscles attached to the arytenoids are responsible for the closure and opening of glottis. The ***prearytenoids*** are found in the depressions situated below the inner surface of upper ends of arytenoids.

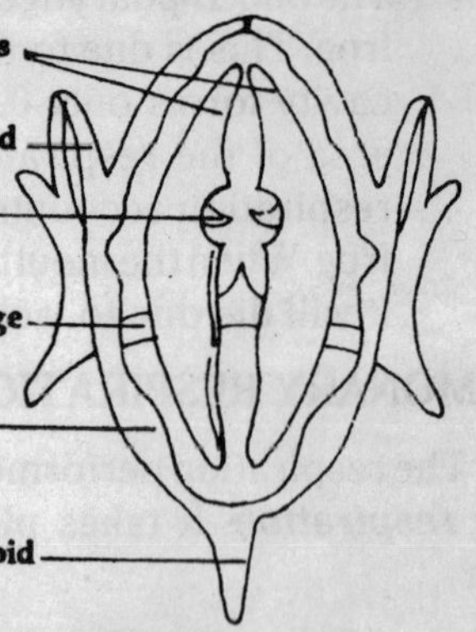

Laryngo-tracheal chamber

The laryngo-tracheal chamber is held in position between the posterior cornua of hyoid apparatus. It opens into the pharynx through a slit like ***glottis*** infront. It also opens into lungs by a pair of openings. The inner lining of laryngo-tracheal chamber is produced into a pair of flaps of epithelium called ***vocal cords.*** The narrow slit present between the ***vocal cords*** is called ***rimaglottidis***. Air passes out through this slit during expiration. Whenever the air is forced out of the lungs the free ends of the lungs the free ends of the vocal cords are vibrated and sound is produced. So, it is called ***voice box***. In male frogs vocal sacs act as resonators to amplify the sound.

A pair of lungs is present on either side of the heart in the thoracic cavity. They are thin walled elastic and oval shaped bag-like structures. Each lung is connected with laryngo tracheal chamber by short inconspicuous bronchus. These are surrounded by ***pleuroperitoneal layers***. The inner layer of the lungs is extended inwards to form into a number of colds called ***primary trabeculae*** which fold further to form ***secondary trabeculae***. The trabeculae divide each lung into a number of small chambers as in honey comb. These small chambers between the secondary trabeculae are called ***alveoli*** which open into large chambers called ***infundibulae***. The alveoli are kept moist with mucous and are richly supplied with blood capillaries by pulmonary artery. Bronchioles are absent. Lungs of frog also serve as ***hydrostatic organs***.

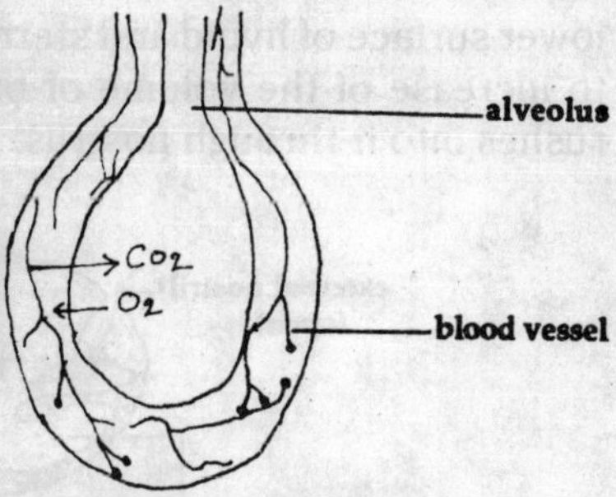

Alveolus enlarged

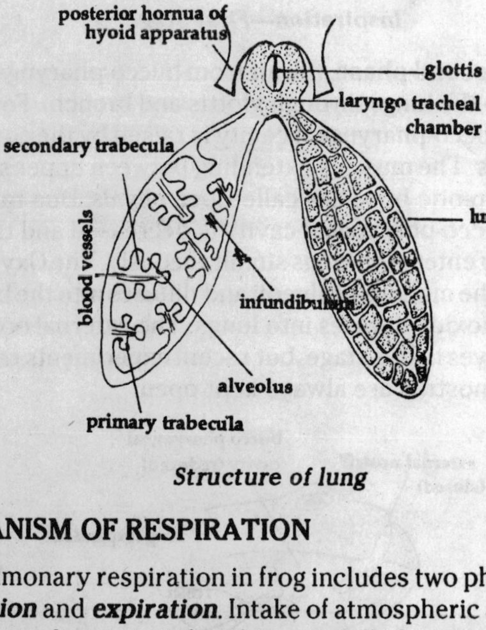

Structure of lung

MECHANISM OF RESPIRATION

Pulmonary respiration in frog includes two phases namely ***inspiration*** and ***expiration***. Intake of atmospheric air is inspiration and sending out air frog lungs is called expiration.

Inspiration

Inspiration is an active process and takes place in two phases. **In the first phase**, the air enters the bucco-pharyngeal cavity through nostrils. To accomplish this process, the mouth and glottis are closed and the nostrils are kept open. The bucco-pharyngeal cavity is enlarged by means of lowering its floor by the contractions of ***sternohyals***. Muscles extending between lower surface of hyoid and sternum are called ***sternohyals***. Due to increase of the volume of buccọ pharyngeal cavity, the air rushes into it through nostrils.

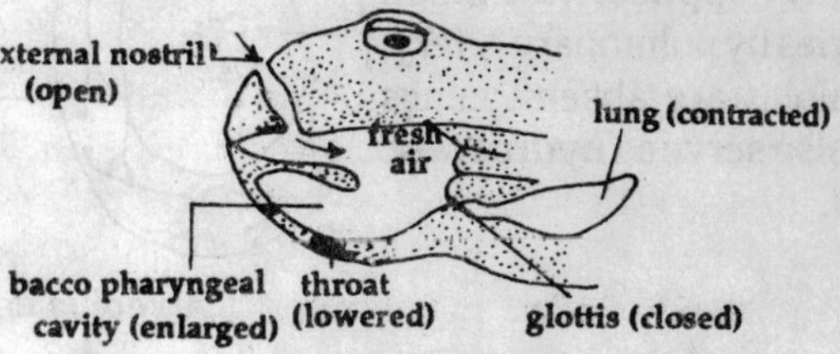

Inspiration—First Stage

In the second phase, the air from bucco-pharyngeal cavity is forced into the lungs through glottis and bronchi. For this the floor of the bucco-pharyngeal cavity is raised by the contraction of ***petrohyals***. The muscles extending between upper surface of hyoid and pro-otic bone are called *petrohyals*. Due to this, the volume of bucco-pharyngeal cavity is decreased and the glottis is opened. Air enters the lungs simultaneously. The Oxygen of air dissolves in the mucous of alveoli and diffuses into the blood and the carbon dioxide diffuses into lungs. The external nostrils are closed by valves in this stage, but recent experiments reveal that the external nostrils are always kept open.

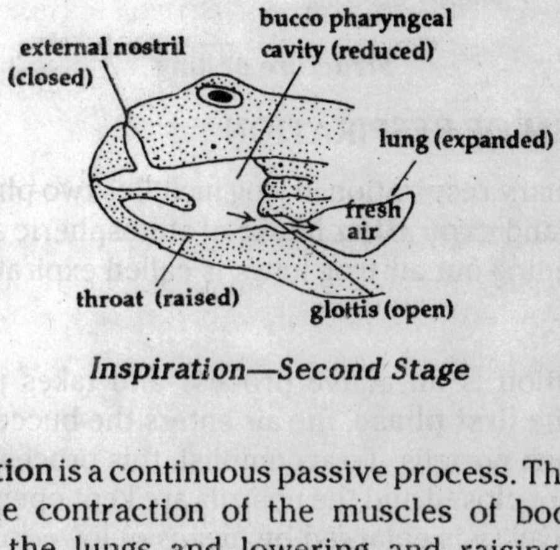

Inspiration—Second Stage

Expiration

Expiration is a continuous passive process. This is brought about by the contraction of the muscles of body wall, the recoiling of the lungs and lowering and raising of floor of buccopharyngeal cavity. It takes place in two stages.

In the first phase, the thoracic muscles contract and increases pressure on lungs. At the same time the floor of the bucco-pharyngeal cavity is also lowered to increase the cavity. So air rushes into the cavity from lungs.

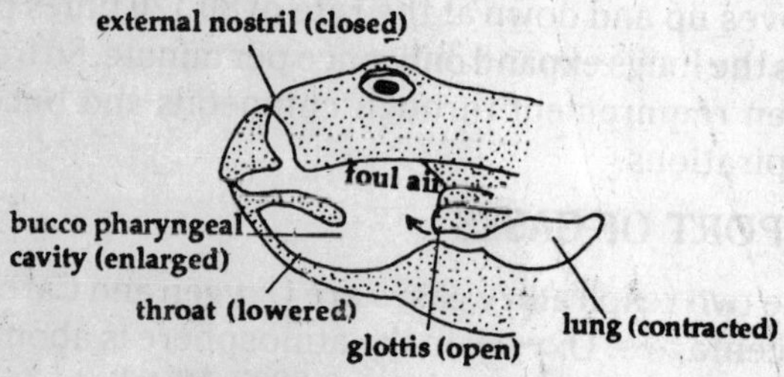

Expiration—First Stage

In the second phase, the glottis is closed, nostrils are opened and floor of the bucco pharyngeal cavity is raised. So the air goes out through nostrils. Pulmonary respiration is important when the frog is out side water. It accounts 64% of the total respiration.

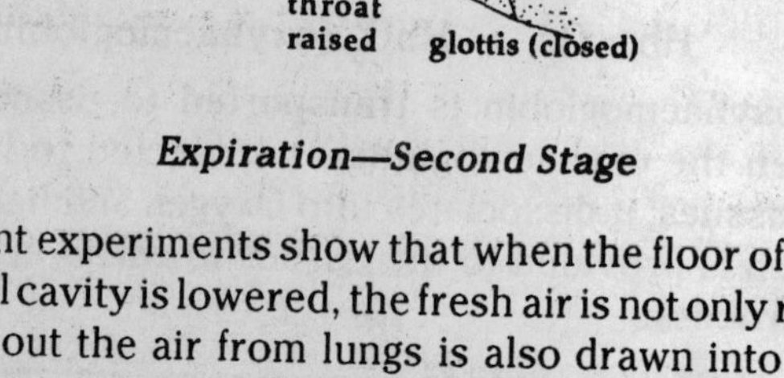

Expiration—Second Stage

Recent experiments show that when the floor of the bucco-pharyngeal cavity is lowered, the fresh air is not only rushed into the cavity out the air from lungs is also drawn into the cavity where the two airs mix forming mixed air. When the floor of bucco-pharyngeal cavity is raised, the mixed air goes outside the body through the external nostrils as well as goes into the lungs through the glottis. It is also important to note that the expiration of air from lungs into the bucco pharyngeal cavity occurs just before the second phase of inspiration.

The pulmonary respiration is not that important in frog. It is indicated by the fact that when the laryngotracheal chamber is tied with a thread, the lung respiration is stopped but the frog has continued its normal life. It is also evident that the buccal floor moves up and down at the rate of 80-120 times per minute, where as the lungs expand only once per minute. So frog receives its oxygen requirement through cutaneous and bucco-pharyngeal respirations.

TRANSPORT OF GASES

The two respiratory gases are Oxygen and Carbondioxide. The percentage of Oxygen in the atmosphere is about 21% while that of Carbon dioxide is about 0.03%. Mainly haemoglobin of blood is responsible for the transport of Oxygen and Carbon dioxide. In frog, hundred milliliters of blood contains about 10gm. of haemoglobin.

Transport of Oxygen

The amount of Oxygen transported in dissolved state in plasma is negligible. The principle Oxygen carrier in all the vertebrates is the iron containing reddish respiratory pigment called ***haemoglobin***. It is a conjugated protein. It is present in erythrocytes in all vertebrates.

The haemoglobin combines with Oxygen to form ***oxyhaemoglobin*** in respiratory surfaces having high Oxygen tension. This process is called ***oxygenation***.

$$Hb + O_2 \rightarrow HbO_2 \text{ (oxyhaemoglobin)}$$

The oxyhaemoglobin is transported to tissues through blood. When the oxyhaemoglobin is subjected to low oxygen tension in tissues, it dissociates into Oxygen and haemoglobin. Thus oxygen is provided to the tissues to carry on the energy releasing process.

Transport of Carbon dioxide

CO_2 is transported both in plasma and RBC It is transported either as carbonic acid or carbamino compounds or bicarbonates. About 5% of Carbon dioxide is transported as carbonic acid in dissolved state in plasma.

$$CO_2 + H_2O \rightarrow H_2CO_3 \text{ (carbonic acid)}$$

About 15% of carbon dioxide is transported as carbamino-compounds. The remaining 80% of carbon dioxide is transported as bicarbonates.

The Carbon dioxide diffuses into the red blood corpuscles and quickly forms unstable carbonic acid. The reaction is catalyzed by ***carbonic enhydrase***. This carbonic acid soon dissociates into hydrogen and bicarbonate ions.

$$CO_2 + H_2O \rightarrow \quad H_2CO_3 \rightarrow \quad H^+ + HCO_3$$

The bicarbonates diffuse out into the plasma and combine with sodium to form sodium bicarbonates. The ***sodium bicarbonate*** functions as a buffer in blood. Some amount of carbon dioxide is carried to lungs RBC as ***carbaminohaemoglobin***. When the carbamino-haemoglobin comes in contact with high Oxygen partial pressure in lungs, it dissociates into Carbon dioxide and haemoglobin. Carbon dioxide diffuses into the alveolar air from blood capillaries.

8

Nervous System

The system that is responsible for receiving and responding to various stimuli is ***nervous system***. All living organisms respond to external stimuli. In higher animals the responses are very quich due to the presence of integrated systems namely ***nervous system*** and ***endocrine system***. The study of integrated systems is known as ***"cybernetics."***

The activities of various parts of the body are controlled and coordinated by these two systems. Nervous system plays a dominant role as it shows quick responses to stimuli. It receives stimuli through receptors or sense organs and transmits them to brain, the part of central nervous system. It sends orders to different parts of the body for necessary action.

The nervous system of frog is divisible into 3 parts :

1. **Central nervous system** : It consists of brain and spinal cord.
2. **Peripheral nervous system** : It includes the nerves arising from brain (cranial nerves) and spinal cord (spinal nerves).
3. **Autonomous nervous system or visceral nervous system** : It comprises sympathetic and parasympathetic nervous systems.

CENTRAL NERVOUS SYSTEM

The central nervous system is developed from embryonic

"neural ectoderm." In embryo, the mind dorsal ectodermal cells are changed to a plate called "medullary plate". The centre of the plate deepens. Hence the plate is converted into a ***"medullary groove"***. The edges of the groove unite to produce a tube called ***"neural tube"***. The anterior part of the tube develops into brain and poserior part develops into ***"spinal cord"***.

BRAIN

The brain of frog is a bilaterally cylindrical structure. It lies in the cranium of skull. The brain and spinal cord are covered and guarded by two connective tissue membranes called ***meninges.*** The inner membrane is thin pigmented, highly vascular and closely invests the brain. This is known as ***piamater***. The outer membrane is tough fibrous and lines the interior of the cranial cavity. This is called ***duramater***. The space between piamater and duramater is known as ***subdural space*** which is filled with a kind of lymph called ***cerebrospinal fluid***. It protects the brain and spinal cord from sudden shocks. The ventricles of the brain are also filled with this fluid.

The brain of frog is divisible into 3 main parts.

1. ***Fore brain*** or ***prosencephalon***
2. ***Mid brain*** or ***mesencephalon***
3. ***Hind brain*** or ***rhombencephalon***

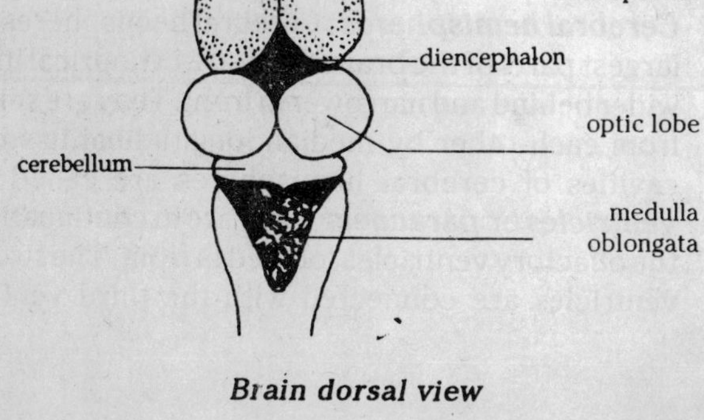

Brain dorsal view

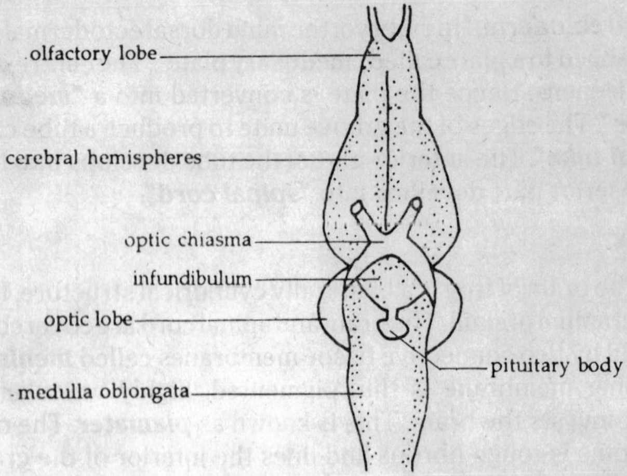

Brain ventral view

1. Fore Brain

It is the largest part of the brain. The fore brain is divided again into two parts, namely (a) ***telencephalon*** (b) ***diencephalon.***

a. Telencephalon : The telencephalon consists of paired ***olfactory lobes*** and ***cerebral hemispheres.***

Olfactory lobes : The olfactory lobes are a pair of large swollen lobes situated at the anterior end of brain. They are fused in the median plane and a median fissure is formed between them. The cavities present in them are called ***olfactory ventricles*** or ***rhinocoels***. The olfactory nerves arise from the anterior ends of the olfactory lobes.

Cerebral hemispheres : Cerebral hemispheres are the largest parts of the brain. They are cylindrical in shape, wider behind and narrower in front. They are separated from each other by median longitudinal fissure. The cavities of cerebral hemispheres are called ***lateral ventricles*** or ***paracoels***. They are in continuation with the olfactory ventricles located in front. The two lateral ventricles are connected with the third ventricle of diencephalon through an opening called ***foramen of Monro***. The thick roof of the cerebral hemispheres is called ***pallium***. It contains less number of neurons. The ventro lateral walls of cerebral hemispheres are thick by many nerve fibres or nerve tracts called ***"corpora striata"***. The corpora striata of the two sides are joined by a pair of transverse tracts of fibres called ***anterior commissure.***

b. Diencephalon : The diencephalon or thalamencephalon is the unpaired part of fore brain lying behind the cerebral hemispheres and infront of the mid brain. The narrow cavity present in it is called ***diocoel*** or ***third ventricle***. The roof of diocoel is then, non nervous and vascular. It is called ***anterior choroid plexus***. The thickened lateral walls of the diencephalon are known as ***optic thalami.***

A hollow, swollen outgrowth called pineal stalk is situated behind the anterior choroid plexus on diencephalon. The pineal stalk has a rounded ***pineal body*** or ***epiphysis*** at its top. In the tadpole stage, penial body and penial stalk are joined with each other but in adult frog the pineal body gets separated from pineal stalk. The pineal body comes out of the skull and it is situated below the skin as ***brow spot***.

The floor of the diocoel is called hypothalamus. On the ventral surface of ***hypothalamus*** lies the crossing of two optic nerves coming from the eyes. This crossing (x-shape) is known as ***optic chiasma***. Behind this crossing, lies a hallow, lobed process called ***infundibulum***. A flattened, oval body, the hypophysis or ***Rathkes pouch*** (pituitary gland). The infundibulum and hypophysis together form the ***pituitary body***.

2. Mid Brain

It is well developed consisting of two large rounded structures called ***optic lobes*** or ***corpora bigemina***. The two lobes are connected by a transverse band of nerve fibres called ***posterior commissure***. The optic lobes are situated behind diencephalon and infront of cerebellum. These optic lobes are hollow and the cavities are called ***optocoels*** or ***optic ventricles***. The floor of the mid brain is thickened by fibrous tracts called ***crura cerebri.*** These fibers extend from medulla to cerebral hemispheres. The narrow duct which connects 3rd ventricle (diocoel) with the 4th ventricle (metacoel) is known as ***"Iter"*** or ***"aqueduct of sylvius"***. It passes between the two optic lobes.

3. Hind Brain

The hind brain includes 2 parts :

a. ***Metencephalon*** or ***cerebellum*** and

b. ***Myelencephalon*** or ***medulla oblongata.***

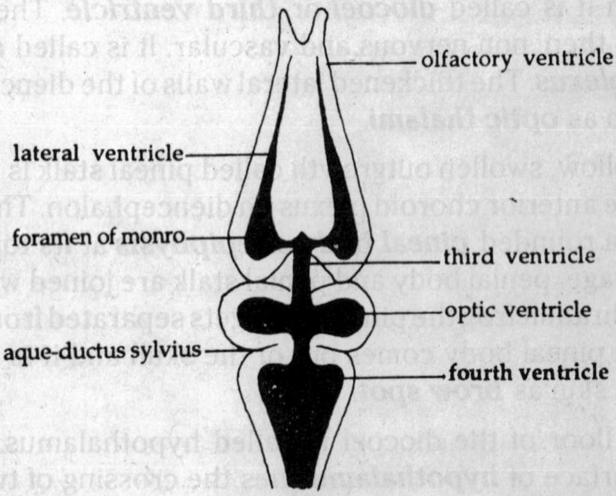

Ventricles of Brain

a. Metencephalon (cerebellum) : It is a single solid transverse ridge-like structure lying dorsally behind the optic lobes. It is poorly developed in frog.

b. Myelencephalon (medulla oblongata) : It is the posterior part of brain. It is a short conical structure which continues behind as spinal cord. It contains a trinagular cavity known as ***metacoel*** or ***fourth ventricle***. It is connected with the cavity of diencephalon through ***iter***. The roof of metacoel is thrown into non-nervous highly vascular folds to form the ***posterior choroid plexus***. The medulla oblongata can be treated as anterior expanded part of spinal cord. It is well developed and an important part of the brain.

BRAIN

- Fore brain (Prosencephalon)
  - Telencephalon
  - Diencephalon Thalamencephalon
- Mid brain (Mesencephalon)
- Hind brain (Rhombencephalon)
  - Metencephalon
  - Myelencephalon

FUNCTIONS OF BRAIN

Brain is the centre for control and coordination of all vital activities. It receives impulses from different parts of body through sensory nerves and sends orders through motor nerves to different parts of the body for necessary action.

The functions of various parts of the brain are given below:

- The olfactory lobes are the centres of sense of smell.
- Cerebral hemisphers are the centres of consciousness, intelligence, memory and voluntary actions.
- Diencephalon of the brain controls metabolism of fats and carbohydrates. It also regulates genital functions and sleep. It plays a role in perceiving the sense of vision. ***Removal of this part of brain causes blindness.***
- Optic lobes are concerned with sense of sight. They also control the movement of eye muscles.
- The cerebellum serves for coordinating muscular actions and maintaining balance during movement.
- Medulla oblongata is an important part of the brain. It controls involuntary functions such as heart beat, respiration, swallowing, taste, hearing and sound production. ***Removal of medulla oblongata causes the death of the animal.***

Spinal Cord

The spinal cord is the second part of central nervous system. The medulla oblongata is continued behind into a spinal cord. It comes out of foramen magnum and runs through neural canal of vertebral column. Just like brain the spinal cord is also covered by two meninges *viz.*, ***outer duramater*** and ***inner piamater.***

In frog, the spinal cord is elongated, more or less cylindrical, bilaterally symmetrical and slightly flattened in dorsoventral direction. The spinal cord terminates into a tapering filament called ***filum terminale*** which lies in the urostyle. The spinal cord is swollen at the level of fore limbs and hind limbs to form two conspicuous enlargements. Anterior enlargement of the spinal cord present at the level of fore limbs is called ***brachial enlargement***, while the posterior enlargement lying between hind limbs is called sciatic ***enlargement***.

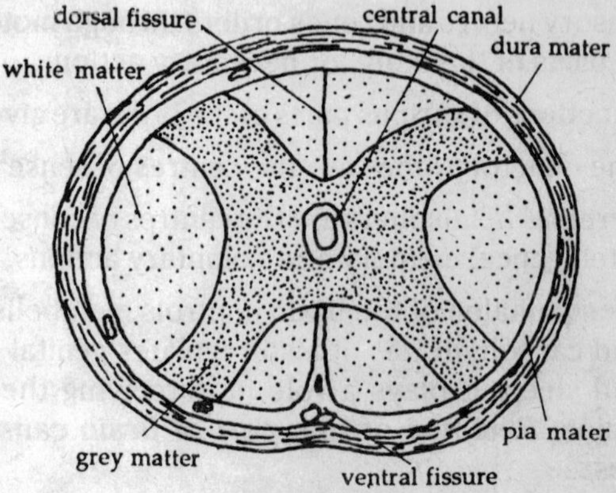

T.S. of Spinal cord

There are two longitudinal fissures in mid dorsal and mid ventral regions of spinal cord. They are known as ***dorsal fissure*** and ***ventral fissure***. The ventral fissure is more conspicuous than the dorsal fissure and it contains blood vessels. The fissures divide the spinal cord into right and left halves.

In the centre of the spinal cord there is a canal called ***central canal***. This canal is lined by single layer of ciliated epithelial cells. Anteriorly, it is in continuation with the cavities of the brain and ends blindly at the posterior end. Surrounding central canal there is a ***grey matter***. It consists of nerve cell bodies and proximal parts of axons. The corners of the grey matter is produced into a pair of ***dorsal*** and a pair of ***ventral horns.***

The grey matter is surrounded by ***white matter***. The white matter consists of ***myelinated parts of axons***. The white matter is divided into four zones, *viz.*, ***a dorsal funiculus***, ***ventral funicules*** and two ***lateral funiculi***. This zonation is formed due to the horns of the grey matter stretched into the white matter. ***The dorsal half of the spinal cord is sensory and ventral half is motor in function.***

PERIPHERAL NERVOUS SYSTEM

The peripheral nervous system includes ***cranial nerves*** and ***spinal nerves.***

The nerves which arise from brain are **cranial nerves** and that of spinal cord are **spinal nerves**. The cranial nerves come out of cranial cavity through ***cranial foramina***. Spinal nerves come out of the neural canal through the intervertebral foramina. The nerves of the peripheral nervous system innervate different parts of the body.

Based on the function, the nerves are of 3 types.

a. Sensory nerves : The sensory nerves are also known as ***afferent nerves***. These nerves carry impulses from sense organs like nose, tongue, skin etc. to the central nervous system.

b. Motor nerves : These nerves are also known as ***efferent nerves***. These nerves carry orders from central nervous system to the effectors (muscles and glands).

c. Mixed nerves : These nerves conduct impulses in both directions. They contain both afferent and efferent fibres.

CRANIAL NERVES

In frog there are 10 pairs of cranial nerves.

I. **Olfactory nerve** : This is a ***sensory nerve*** arising from the anterior end of the olfactory lobe and innervates neurosensory cells of ***olfactory chamber***. Olfactory impulses are conducted to the brain from nasal sac through this nerve.

II. **Optic nerve** : It is a ***sensory nerve*** arising from the lateral side of the optic lobe. The nerve innervates ***retina of the eye***. Visual impulses are carried to the brain from retina through this nerve.

III. **Occulomotor nerve** : This is a ***motor nerve*** arising from cruracerebri of mid brain. It innervates the ***internal superior rectus***, ***inferior rectus and anterior rectus*** and ***inferior oblique muscles*** of eye ball.

IV. **Trochlear or pathetic nerve** : Trochlear nerve is a small ***motor nerve*** arising from the dorsal side of the brain between optic lobes and cerebellum. This nerve innervates the ***superior oblique muscle*** of eye ball.

V. **Trigeminal nerve** : This is a ***mixed nerve*** arising from anterolateral parts of medulla oblongata. At its base, it bears a ganglion called ***gasserian ganglion***. It divides into 3 branches.

a. **Opthalmic nerve** : It passes along the dorsal border of the orbit (between the skull and the eye ball) and innervates the ***skin of snout. It is sensory in nature.***

b. **Maxillary nerve** : It runs below the eye ball and supplies to the skin of upper jaw. It is also ***sensory*** in nature.

c. **Mandibular nerve** : It runs along the outerside of the ***lower jaw*** and innervates ***muscles and skin***. It is mixed in nature.

VI. **Abducens nerve** : It is a small ***motor nerve*** arising from the ventral side of medulla oblongata and innervates the ***external rectus*** and ***retractor bulbi muscles*** of eye ball.

VII. **Facial nerve** : it is a ***mixed nerve*** arising from lateral sides of the medulla oblongata just behind the 5th cranial nerve. It joins with ***gasserian ganglion*** before leaving the skull. It is divided into two branches.

a. **Palatine nerve** : It runs forward below the orbit and innervates mucous membrane (roof) of the buccal cavity. It is ***snesory*** in nature.

b. **Hyomandibular nerve** : It crosses over columella

auris of the middle ear and innervates the muscles of the lower ***jaw*** and ***hyoid apparatus***. It is ***motor*** in nature.

VIII. **Auditory or accoustic nerve** : This is a ***sensory nerve*** arising from the lateral sides of medulla and innervates the sensory cells of ***internal ear***.

IX. **Glossopharyngeal nerve** : This is a ***mixed nerve*** arising from the lateral sides of medulla behind auditory nerve. It innervates mucous membrane of the tongue, pharynx and muscles of hyoid. It gives out a delicate branch which joins hyomandibular branch of the seventh cranial nerve.

X. **Vagus or pneumogastric nerve** : This is the largest ***mixed nerve*** arising from the lateral sides of the medulla oblongata. It has common origin with the nith nerve. After emergin from skull, the nerve swells and forms a ***vagus ganglion***. The vagus nerve is divided into five branches.

a. **Cutaneous nerve** : It is ***sensory nerve*** innervating the dorsal body wall.

b. **Cardiac nerve** : It is a ***motor nerve*** innervating the heart.

c. **Laryngeal nerve** : It is a ***mixed nerve*** going to the laryngo-tracheal chamber.

d. **Pulmonary nerve** : It is a ***motor nerve*** going to the lungs.

e. **Gastric nerve** : It is a ***mixed nerve*** inneravting the stomach.

SPINAL NERVES

The nerves arising from the spinal cord are called ***spinal nerves***. There are 10 *pairs of spinal nerves* in frog, but in ***Rana tigrina***, only 9 pairs are found. Each spinal nerve arises from the spinal cord by two roots, ***a dorsal or sensory root and a ventral or motor root***. These roots contains a small dorsal ganglion just outside the neural canal. After originating from the spinal cord thus formed passes out of neural canal through intervertebral

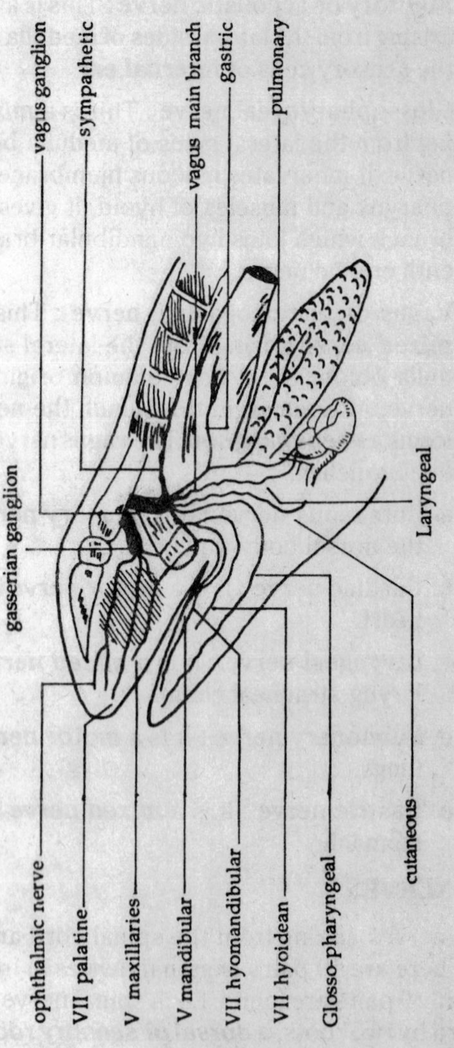

Cranial nerves

Cranial Nerves

Sl. No.	Name of the Cranial nerve	Branches	Nature	Origin	Distribution
1.	Olfactory nerve	———	Purely somatic sensory nerve.	Anterior end of olfactory lobe.	Mucous membrane of the nasal sacs.
2.	Optic nerve	———	Purely somatic sensory nerve.	Ventro-lateral part of the optic lobe	Retinal of they eye.
3.	Occulomotor nerve	———	Somatic motor nerve.	Ventral surface of the midbrain (crura cerebri)	Muscles of the eye (internal, superior rectus, inferior and superior oblique-muscles)
4.	Trochlear nerve (Pathetic)	———	Somatic motor nerve.	Dorsally between the optic lobes and and cerebellum)	Fifth muscle of the eye (superior oblique muscle)
5.	Trigeminal nerve		Mixed nerve	Antero-lateral part of medulla oblongata	
		a. Opthalmic	Somatic sensory nerve		a. Skin of snout
		b. Maxillary	Somatic sensory nerve		b. Skin of the upper jaw and lower eye lid.
		c. Mandibular	mixed nerve.		c. Muscles of the lower jaw and its skin.
6.	Abducens nerve	———	Somatic motor nerve	Ventral side of the medulla oblongata	Muscles of the eye

(Contd..)

Cranial Nerves (Contd.)

Sl. No.	Name of the Cranial nerve	Branches	Nature	Origin	Distribution
7.	Facial nerve		Mixed nerve	Lateral parts of the medulla oblongata	Mucous membrane (roof of the buccal cavity)
		a. Palatine	Visceral sensory nerve		
		b. Hyomandibular	Visceral motor nerve	————	Muscles of the lower jaw and hyoid apparattus.
8.	Auditory nerve	————	Somatic sensory nerve	Lateral parts of medulla oblongata	Internal ear (auditory organ)
9.	Glosso-pharyngeal	Sends a branch to the facial nerve.	Mixed nerve (Visceral sensory and visceral motor).	Lateral parts of medulla oblongata.	Mucous membrance of tongue, pharynx and muscles of the hyoid.
10.	Vagus nerve	Mixed nerve (Somatic, visceral sensory and visceral motor)	Lateral parts of medulla oblongata		
		a. Cutaneous	Sensory nerve	————	Skin of the dorsal body wall
		b. Cardiac	Motor nerve	————	Heart
		c. Laryngeal	Mixed nerve	————	Laryngo-tracheal chamber
		d. Pulmonary	Motor nerve	————	Lungs
		e. Gastric	Mixed nerve	————	Stomach

foramen. All the nerves at their origin are covered by silvery white galnds called ***swammerdam or periganglionic glands.***

These glands are lymphatic in origin and serve as reservoirs of calcium, which supply calcium whenever the nerves require.

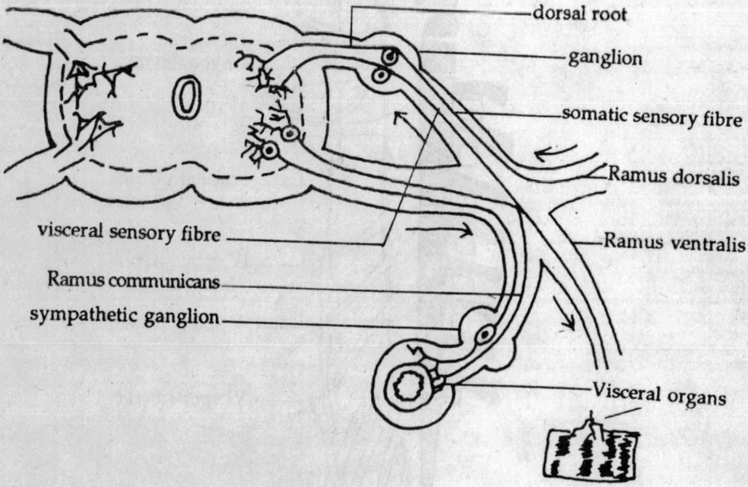

Spinal nerve formation

Each spinal nerve divides into 3 branches. These are :

a. **Ramus dorsalis** (dorsal branch) : It divides into several branches which enter the skin and muscles of dorsal body wall. It contains ***somatic sensory fibres.***

b. **Ramus ventralis** (ventral branch) : It is a long main branch. It supplies nerves to the muscles of belly (Ventral body wall). It contains ***somatic motor fibres.*** The prominent ventral branch is generally considered as the spinal nerve.

c. **Ramus communicans** (visceral branch) : It joins with sympathetic ganglion. It contains both ***visceral sensory and visceral motor fibres.***

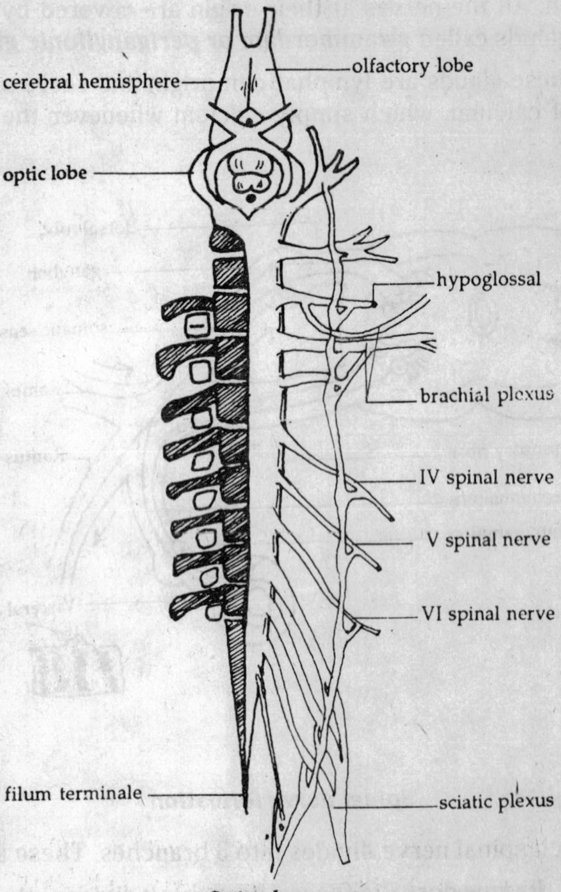

Spinal nerves

The **first spinal nerve** is called ***hypoglossal***. It comes out between first and second vertebrae. It runs forward along the floor of buccal cavity and innervates tongue and muscles of the hyoid.

The **second spinal nerve** is known as ***brachial nerve***. It is a stout nerve emerging between second and third vertebrae. The second, third and a branch of first spinal nerve join to form a nerve net, called ***"brachial plexus"***. After forming brachial plexus the second spinal nerve passes into the fore limb to innervate its muscles.

The **third spinal nerve** is a small nerve emerging between third and the fourth vertebrae. It immediately joins second spinal nerve forming brachial plexus and then separates itself and innervates muscles of the back.

The **fourth, fith** and **sixth spinal nerves** are slender nerves emerging out in between 4/5, 5/6, and 6/7 vertebrae. They run obliquely backwards and innervate muscles and skin of the ventral body wall.

The **seventh, eighth** and **ninth spinal nerves** are large and run almost backwards. They unite with one another to form a network known as ***sciatic plexus***. The eighth and ninth nerves unite a continue in the hind limb as ***sciatic nerve***.

The tenth spinal nerve is called ***coccygeal nerve***. It is a small nerve which emerges out through a small foramen on the side of urostyle near its anterior end. It is frequently present only on one side. It may be absent altogether in certain cases. When present, it innervates the cloaca, urinary bladder and other nearby parts.

NERVES

- **Cranial nerves**
  - **Sensory nerves**
    - **Olfactory (1st)**
    - **Optic (2nd)**
    - **Auditory (8th)**
  - **Motor nerves**
    - **Occulomotor**
    - **Trochlear**
    - **Abducens**
  - **Mixed nerves**
    - **Trigeminal**
    - **Facial**
    - **Glassophary-ngeal**
    - **Vagus nerves**
- **Spinal nerves**

Classification of nerves in Frog

AUTONOMOUS NERVOUS SYSTEM

The name ***autonomous*** (Gr. autos = self, nomos = governing) indicates that it is a self governing nervous system. It is a partly independent system. It is not under the voluntary control of nervous system. The activities of some of the organs which are not under the control of the will of an organism are controlled by this system. The autonomic nervous system controls activities like heart beat, peristalsis in alimentary canal, contraction of urinary bladder secretion of digestive juices from digestive glands, etc. The autonomous nervous system is also known as ***"visceral nervous system"***.

The central nervous system generally, regulates various visceral organs and movements of involuntary muscles. Though it is involuntarily controlled by the nerve centres located in the central nervous system, it has connection with spinal nerves and some cranial nerves.

The autonomous nervous system consists of ***visceral sensory*** and ***visceral motor fibres***. It has two divisions : 1. ***sympathetic nervous system*** and 2. ***para-sympathetic nervous system***. They ensure smooth functioning of the various organs.

SYMPATHETIC NERVOUS SYSTEM

It consists of a pair of slender longitudinal chains of ganglia, which lie on either side of the dorsal aorta. These chains are called ***sympathetic nerve chains*** or ***cords***. They continue forward along the outer sides of systemic arches. Anteriorly, each sympathetic nerve cord enters the skull through ***jugular foramen*** and finally ends in ***vagus*** or ***gasserian ganglion***. The sympathetic nerve chain is connected with corresponding spinal nerve by a narrow branch called ***ramus communicans***.

Several nerves arise from sympathetic ganglia, which innervate various visceral organs like stomach, intestine, spleen, kidneys, urinary bladder, gonads etc. The nerve endings of the sympathetic nervous system secrete a chemical substance called ***sympathin*** which is functionally similar to ***adrenalin***. This sympathin has an accelerating effect on the activities of various visceral organs.

PARASYMPATHETIC NERVOUS SYSTEM

This is a functionally different component of the autonomous nervous system. The nerve fibres of parasympathetic nerves are found in ***third, seventh, ninth and tenth cranial nerves and the spinal nerves of the sacral region.*** The ganglia of the parasympathetics system are located in the wall of heart, stomach, lungs, urinary bladder etc. The nerve endings of parasympathetic system secrete ***acetylcholine.*** This substance has inhibiting effect which retards the over activity of the organs. Thus the normal working of the body is ensured by opposing functions of two components (sympathetic and parasympathetic) of autonomous nervous system.

The opposing functions of sympathetic and parasympathetic systems

Organ & Activities	*Sympathetic control*	*Parasympathetic control*
1. Rate of heart beat	Increases	Decreases
2. Stomach		
a) Tonus and motility	Decrease	a) Increase
b) Secretions	Decrease	b) Increase
3. Intestine Tonus and motility	Decrease	Increase
4. Bronchial muscles	Dilate	Constrict
5. Blood vessels	Constrict (except coronaries and skeletons).	Dilate
6. Urinary bladder	Relaxation of muscles occur	Contraction of the muscles occur
7. Peristalsis of alimentary canal	Decreases	Increases
8. Secretions of salivary and gastric glands	Decrease	Increase

SENSE ORGANS OF FROG

In Frog, various types of receptors or organs of special sense are found. The receptors are supplied with nerve fibres and thus convey the stimulus to the central nervous system. The receptors in Frog are ***Tangoreceptors, olfactory receptors, Gustatory receptors, photo receptors and statoaccoustic receptors.***

The entire skin in Frog serves as organ of touch as it is supplied with sensory nerve endings situated in the space between the cells.

The olfactory or nasal sacs act as olfactory receptors or organs of smell.

Gustatory receptors or organs of taste are found in the form of taste buds which are confined to the tongue and roof of buccal cavity.

EYE

The photoreceptors or organs of sight are eyes. They are situated in the orbital fossae. There are three eyelids in Frog. The upper eye lid is not movable and the lower eye lid is movable.

The transparent membrane covering the eyes when Frog goes inside water or under ground is nictitating membrane. The ***nictitating membrane*** arises from the upper part of lower eye lid and is protective in function.

The movement of nictitating membrane is controlled by two muscles namely ***retractor bulbi***, and ***levator bulbi***. The eye ball is moved by set of six extrencic muscles, out of which four are ***rectus muscles*** and two are ***superior*** and ***inferior oblique muscles.***

The surface of eyes is remained moist due to the presence of ***Harderian glands***, situated at the inner angle of the eye. The eye ball is made of three concentric layers outer ***sclerotic***, middle ***choroid*** and ***inner retina.***

The outermost dark blue layer which is made up of fibrous connective tissue and cartilage is ***sclerotic***.

The sclerotic layer is continued infront as an arched transparent membrane called ***cornea***. The cornea is covered externally by a thin transparent membrane called ***conjunctiva***. The conjunctiva is formed by epidermis.

The choroid layer is made of loose connective tissue fibres having many blood capillaries and pigment cells. The choroid thickens as a ring like ciliary body towards anterior end and continues as **iris**. In the middle of iris is an apenture called *pupil*. The pupil appears like a black spot in the middle of the yellow iris. Light enters the eye through pupil and light is controlled by iris.

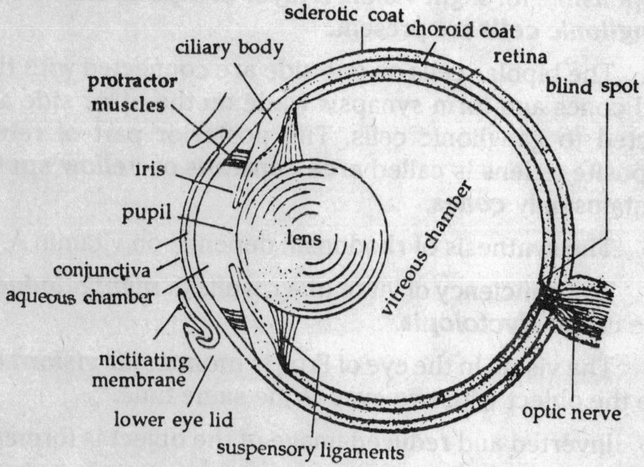

Eye

There is no power of accommodation in Frog. Two sets of involuntary muscles are present. One set is ***sphincter muscles*** arranged in concentric rings around pupil which reduce pupil. The second set of muscles are ***radial muscles*** which dilate pupil.

Immediately behind and in close contact with iris is a transparent ***crystalline lens***, enclosed in lens capsule. The shape of lens is ***biconcave***.

The lens is suspended behind the pupil by a membranous ***suspensory ligament***.

The chamber present behind the lens is ***vitreous chamber*** and in front of the lens is ***aqueous chamber***. The aqueous chamber is filled with salty ***aqueous humour*** and vitreous chamber is filled with gelatinous ***vitreous humour***.

The inner most retinal layer is with two layers, an outer layer of pigment cells and an inner layer of sensory or ***retina proper***. The retina proper is connected to the optic nerve. At the exit of optic nerve there is a ***blind spot***. The light sensitive cells present in the retina are ***rods*** and ***cones***.

The rods are meant to perceive the amount of light while the cones for distinguishing colours. The pigment present in rods is called ***rhodopsin*** and that of cones is ***lodopsin***. Cones are responsible for day light vision and colour vision and rods are responsible for night vision. A layer of ***bipolar nerve cells*** and ***ganglionic cells*** are present.

The bipolar cells on one side are connected with the rods and cones and form synapsis while on the other side are connected to ganglionic cells. The posterior part of retina just opposite to lens is called ***area centralis*** or y***ellow spot*** which contains only ***cones***.

The synthesis of rhodopsin depends on vitamin A.

The deficiency of vitamin A results in ***night blindness***. It is also called ***Nyctolopia***.

The vision in the eye of Frog is ***monocular vision*** i.e. it can see the object with two eyes at the same time.

Inverted and reduced image of the object is formed in the eye.

Rods are more numerous in the eyes of nocturnal animals and cones are poorly developed in these animals.

EAR

The statoaccoustic organs or organs of hearing in frog are ears. The ear of frog includes tympanum middle ear and internal ear and external ear is absent.

The middle ear is an air filled chamber communicating with the pharynx by a slender ***eustachian tube***. The cavity of middle ear is bounded by the ***tympanic membrane*** or ***ear drum***. This is stretched over a ring of cartilage, called ***annulus tympanicus***. It is the modified skin.

Extending across the tympanic cavity is a club shaped rod, ***columella auris***. It is a modified hyomandibular. Columella auris is an ear ossicle in Amphibians, reptiles and Aves.

The internal ear consists of bony auditary capsule. It is formed by the ***prootic bone.*** The auditory capsule is filled with watery fluid called ***perilymph*** in which membranous labyrinth floats.

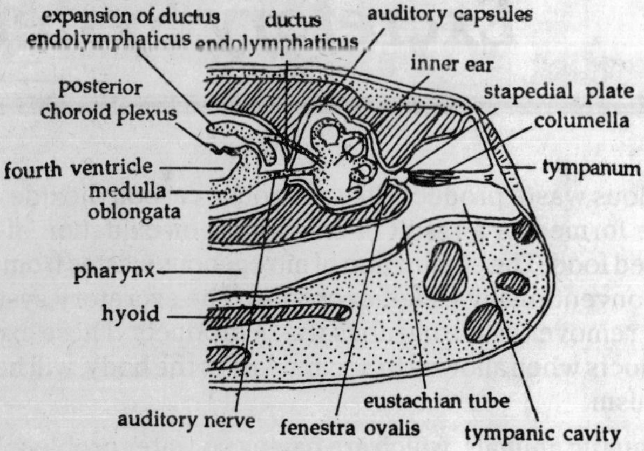

Ear

The membranous labyrinth consists of ***dorsal utriculus, ventral sacculus*** and ***three semicircular canals.*** The entire membranous labyrinth is filled with a fluid called ***endolymph.*** The endolymph contains pieces of $CaCO_3$ forming ***otoliths*** or ***ear stones.***

The one end of semicircular canal is dilated to form a small round ***ampulla.***

From the posterior side of sacculus arise two small dilations called ***lagena*** or ***cochlea*** and ***parabasillaris.*** From the inner side of the sacculus arises a narrow tube, the ***ductus lymphaticus,*** which contain a milky fluid of $CaCO_3$. The function of inner ear is hearing and equilibrium.

9

Excretory System

Various waste products like ammonia, carbon dioxide and water are formed as a result of catabolism or oxidation of the assimilated food. The elimination of nitrogenous wastes from the body is conventionally called ***excretion***. The excretory system serves to remove the nitrogenous waste products. These excretory products when allowed to accumulate in the body, will harm the organism.

In aquatic animals, which are having no water problem, the nitrogenous waste is excreted in the form of ammonia. Such animals are called ***ammonotelic animals***. Ammonia is toxic if allowed to accumulate in the body. In some animals which partly lead terrestrial life, the toxic ammonia is changed into less toxic urea in the liver. Later, it is removed from the body by excretory organs. Such animals like amphibians which excrete their nitrogenous waste in the form of urea are called ***ureotelic animals***.

Some animals like reptiles and birds cannot afford the loss of water from the body even to the extent necessary to eliminate urea. In these animals the nitrogenous waste is converted into less toxic uric acid which is excreted in crystaline form with no loss of water. The animals eliminating the nitrogenous waste in the form of uric acid are known as ***uricotelic animals***. The chief excretory organs, kidneys, help in the removal of nitrogenous waste.

In the vertebrates like frog, the excretory system is closely associated with reproductive system. Therefore, these two systems are collectively described as ***urinogenital system***.

The excretory system in frog consists of a pair of ***kidneys, their ducts and urinary bladder***.

KIDNEYS

Kidneys are the chief excretory organs of frog. These are a pair of elongated, thin, flat, dark red coloured organs situated in the ***sub-vertebral lymph sinus*** in the body cavity on either side of the urostyle. The kidneys are covered ventrally by a thin peritoneum. The outer border of each kidney is slightly convex but the inner border is straight with 2 or 3 notches. A slender orange coloured structure called ***adrenal*** or ***supra renal gland*** of endocrine function is situated on the midventral surface of each kidney. The gonads are found attached anteriorly to the kidneys. Several finger like yellowish fat bodies are situated in-front of gonads. These are called ***corpora adiposa***.

URETERS

The ureters (excretory ducts) arise from the outer edges of both the kidneys, run back-wards and open separately into the dorsal side of cloaca. The ureters of frog do not correspond to the ureters of the higher vertebrates and should therefore, be called ***wolfian*** or ***mesonephric ducts***. In male frog, the ureters act as genital ducts by carrying sperms. As the ureters serve to conduct reproductive elements besides urine, they are called ***urinogenital ducts***.

The kidney receives blood through renal arteries. The blood from each kidney is collected by renal veins and it finally reaches the heart through postcaval vein. In tadpole, the kidneys are ***pronephric***, whereas in adults these are ***mesonephric***. The ciliated depressions present on the ventral surface of the kidneys are called ***nephrostomes***. They are believed to remove waste materials from coelom.

URINARY BLADDER

This is a thin walled, bilobed saclike structure in which the urine is temporarily stored. It develops as an external outgrowth from the ventral wall of cloaca. The wall of the urinary bladder contains smooth muscles, blood vessels and nerves. It opens into the cloaca through an aperture and this aperture is provided with a sphincter. The urine is passed outside through cloacal aperture, when the urinary bladder contracts at regular intervals.

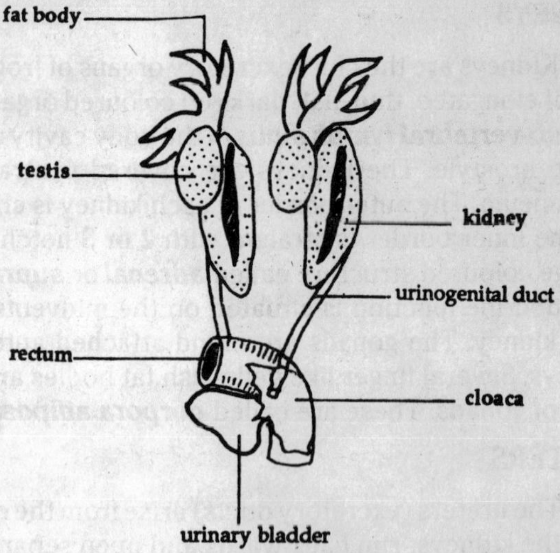

Urinogenital System

INTERNAL STRUCTURE OF KIDNEY

Each Kidney consists of a large number of structural and functional units called ***renal tubules*** or ***nephrons***. These are bounded together by connective tissue and are richly supplied with blood capillaries.

Nephron

Each nephron can be divided into two parts namely, 1. ***malpighian body*** and 2. ***uriniferous tubule.***

Malpighian Body

It consists of thin double walled cup called ***Bowman's capsule*** which encloses as bunch of capillaries called the ***glomerulus***. The glomerulus receives blood from an afferent arteriole of the renal artery. The ***afferent arteriole*** has a large diameter which forms a capillary network (glomerulus) in the cup of the Bowman's capsule. The blood from glomerulus is collected by an ***efferent arteriole*** having smaller diameter than that of the afferent arteriole. The efferent arteriole is formed by the union of capillaries of the glomerulus. The capillary network around the uriniferous tubule is formed by both efferent arteriole and a branch of the renal portal vein. A renal vein is formed by the union of capillaries of the net work.

Uriniferous Tubule

The malpighian body leads into an elongated uriniferous tubule. It is divisible into 4 parts.

- The first part of the uriniferous tubule is a narrow ***neck*** lined with ciliated epithelium.

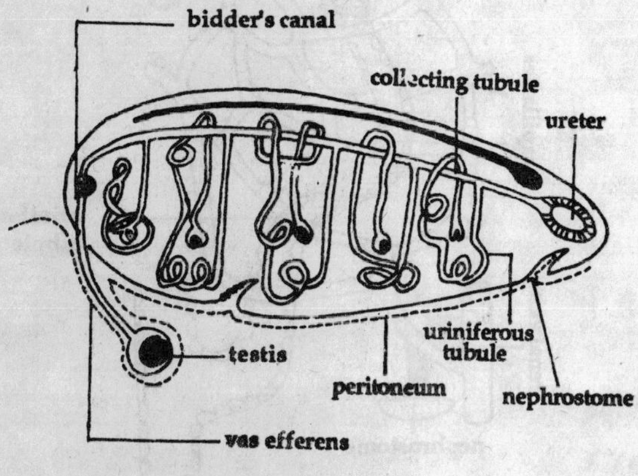

Cross section of Kidney

- The second part is convoluted and it is lined with glandular epithelium.
- The third part is narrow lined with ciliated epithelium.
- The fourth part is also convoluted and is made up of simple columnar epithelium. This distal part opens into transverse ***collecting tubule***. Thus, the uriniferous tubules open into collecting tubule. These tubules run transversely and open into the ureter on the outer margin of the kidney.

Embryonic development of kidney

In the embryo of frog, three pronephric tubules appear first which unite to form into a kidney. It is called ***pronephros***. It is the functional kidney in the tadpole of frog. During metamorphosis, pronephric kidney disintegrates and disappears by the develop-

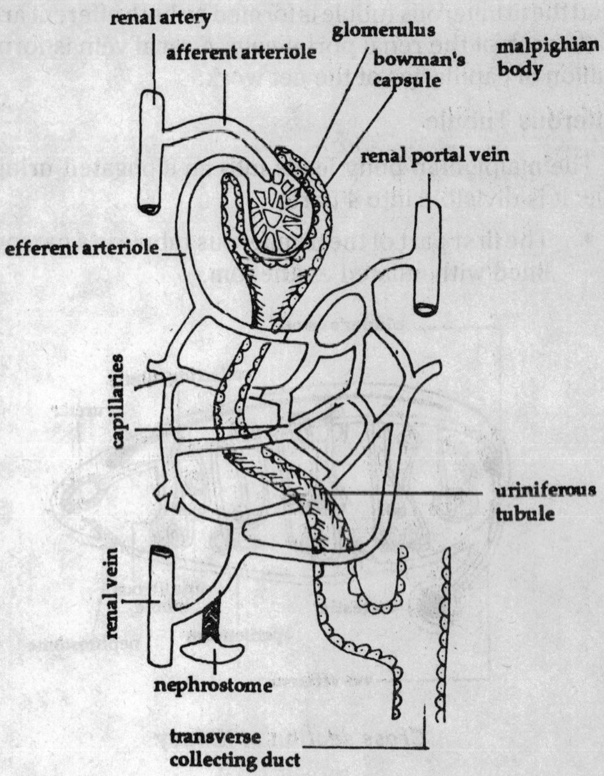

Nephron

ment of mesonephric tubules below the pronephros. These mesonephric tubules unite to form into a kidney called ***mesonephros***. These tubules divide anteriorly and open into coelom through nephrostomes. Posteriorly these tubules unite to form into ***mesonephric duct***. These coiled tubules are connected by the connective tissue and covered by the peritoneum which forms the mesonephric kidney in adult frog.

PROCESS OF EXCRETION

The Process of excretion is completed in three stages viz., ***ultrafiltration, selective reabsorption and tubular secretion.***

Ultrafiltration

The waste substances which are formed during catabolism are brought to the kidneys by the blood through renal arteries and renal portal veins. The blood with waste substances is carried to glomerulus by afferent arteriole. The blood from glomerulus is collected by efferent arteriole. The afferent arteriole is of a larger diameter than the efferent arteriole as mentioned earlier. This difference is diameter results in the increase of hydrostatic pressure in the glomerulus. This pressure causes filtration of blood. This filtration of blood in glomerulus under great pressure is called ***ultrafiltration***.

During this process, the macromolecules like the blood corpuscles, blood proteins and fats are retained in the blood as they cannot penetrate through the wall of glomerulus. But the substances like urea, glucose large quantities of water sulphates, chlorides, phosphates of Sodium, Potassium, Calcium and Magnesium are filtered through the glomerulus into the cavity of the Bowman's capsule. The fluid that is collected into the cavity of the Bowman's capsule is called ***glomerular filtrate or primary urine.***

Selective reabsorption

While the primary urine is passing through the convoluted tubule, the substances like glucose, aminoacids, inorganic chlorides and some water are reabsorbed from the filtrate into the blood of capillaries. This process is called ***selective reabsorption.***

Tabular Secretion

Uriniferous tubule secretes wastes like urea and unnecessary salts from the blood of capillary network into the filtrate. This process is known as tubular secretion. After selective reabsorption and ***tubular secretion.*** After selective reabsorption and tubular secretion, the wastes along with water form urine. Now the urine goes into ureters through collecting tubules from where it passes down into the urinary bladder. The urine is temporarily stored in urinary bladder. When the bladder is with full of urine its involuntary muscles contract, then the urine is forced into the cloaca and later it is passed out through the cloacal aperture at intervals.

The glomerular filtrate collected into the cavity of the Bowman's capsule is of lesser concentration than that of urine that is collected in the urinary bladder.

10

Endocrine System

The glands are secretory in function. The glands are of two types. viz., (1) exocrine and (2) endocrine glands. The secretions of endocrine glands are called ***hormones***. The hormones do not pass through ducts, hence the glands are also called ***ductless glands***. But hormones are distributed to the various organs through the blood. Hence such glands are termed as ***"endocrine glands"*** [endo-internal; crinin-secretion].

The various system are co-ordinated by nervous and endocrine system. These two systems together called ***integrated system***. The endocrine system can not influence the organs rapidly as the nervous system does.

The term hormone was proposed by ***Bayliss*** and ***Starling***. Hormones are biocatalysts, which are required in small amounts. These do not participate in biochemical reactions. Hence hormones are also called ***automones*** or ***autocoids***. These are secreted in organs, but their action is else where i.e. normally other than the endocrine glands. For example, the follicle stimulating hormone of pituitary acts on gonads. The hormones carry informations or messages to the ***target organs*** [target organ=organ which is subjected to the action of hormone].

Hormones regulate a variety of physiological activities, like growth, reproduction, temperature, sexual dimorphism etc. Hormones mainly maintain the internal environment or ***homeostasis*** by messages. Hence hormones are also called ***informational molecules*** or ***chemical messengers.***

The study of endocrine glands and their hormones is called ***"endocrinology"***. All hormones are not always proteins. Based on the components, hormones are classified into 4 types.

(1) Protein or peptide hormones: These are made up of proteins or peptides e.g. hormones of hypothalamus, pituitary, parathyroid, pancreas and relaxin of ovary.

(2) Steroid hormones: These are made up of steroids e.g. hormones of adrenal cortex, ovaries and testes.

The names of steroid hormones contain the word ***"stero"***. These stimulate the body organs highly. Hence synthetic steroids are being misused by athelets.

(3) Aminoacid derivatives of biogenic amines: These hormones are derived from aminoacids e.g. melatonin and adrenalin of medulla.

(4) Iodinated aminoacid: e.g. thyroxine.

The different types of endocrine glands in frog are:

(1) Pineal body	(2) Pituitary	(3) Thymus
(4) Thyroid	(5) Parathyroid	(6) Adrenal
(7) Pancreas	(8) Gonads	(9) Mucosa of alimentary canal.

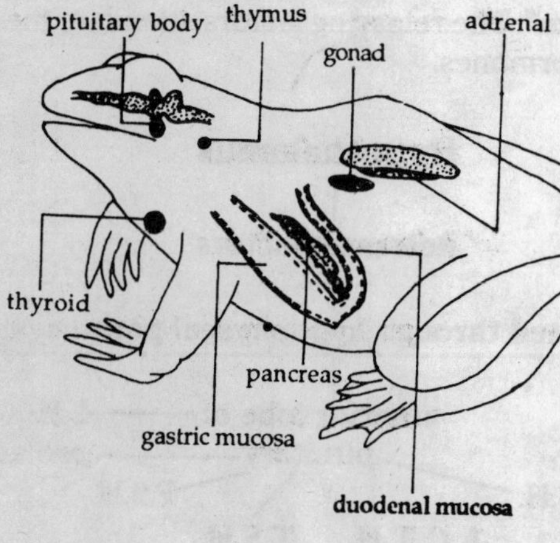

Endocrine glands

PINEAL BODY

It is attached to the dorsal wall of diencephalon by pineal stalk in tadpole larva of frog. Based on its position it is also called ***"epiphysis"*** [Epi-upon; physis-brain]. In adult frog, the pineal gland becomes separated from the stalk, which finally repre-sented as vestigial ***"brow spot"***. The gland secretes the hormone ***"melatonin"***, a biogenic amine in tad pole larva. It gives protec-tive coloration. It makes the skin of larva more pale.

Till recently, it was believed that the pineal gland found in man was a vestigial organ. It is now known that the pineal body secretes melatonin, which is an antigonadial as it may delay the sexual development.

PITUITARY GLAND

The pituitary is small pea-shaped gland situated below the hypothalamus [ventral wall of diencephalon]. It is connected to the hypothalamus by a stalk called ***"infundibulum"***. Based on the position the gland is also called ***"hypophysis"*** [hypo-below; physis-brain]. Until recently, the pituitary gland was considered to be the ***"master of chief gland of endocrine system"***. It is now known that the pituitary itself depends upon the hormones of hypothalamus. The hypothalamus secretes many hormones called ***"neurohormones"*** or ***"releasing factors"***. These are trans-ported to the anterior lobe of pituitary through the "***hypophyse-al portal system***". The releasing factors stimulate the pituitary to secrete its hormones.

Hypothalamus

Releasing factors

Transported through hypophyseal portal system

anterior lobe of pituitary — L.H., prolactin, F.S.H., T.S.H., A.C.T.H., S.T.H.

The pituitary is divided into 3 parts viz., anterior, intermediate and posterior. The anterior and posterior parts of pituitary together called ***"adenohypophysis"***. Where as the posterior part is called ***"neurohypophysis"***.

I. **Anterior lobe of pituitary** : It secretes six hormones, namely A.C.T.H, T.S.H., S.T.H., F.S.H., L.H and prolactin.

1. **A.C.T.H. [Adreno cortico tropic hormone]**: it stimulates the cortex of adrenal gland to secrete gluco, mineral and sex corticoids.
2. **T.S.H. [Thyroid stimulating hormone] or T.T.H [Thyro tropic hormone]**: It stimulates the thyroid gland to secrete thyroxine.
3. **S.T.H. [Somato - tropic hormone] or G.H. [growth hormone]**: It stimulates the growth tissues especially long bones and muscles. The growth hormone is probably the only anterior pituitary hormone which does not stimulate any other endocrine gland.

   G.T.H. [Gonado tropic hormone] or gonadotropins : These are three types (1) F.S.H. (2) L.H. (3) Prolactin [L.T.H.].

(4) **F.S.H. [Follicle stimulating hormone]**: It stimulates the testes in the male to produce sperms and the ovaries in the female to produce ova.

(5) **C.S.H. [Interstitial cell stimulating hormone] or L.H.**: It stimulates the testis to secrete the male sex hormones called ***"testosterones or androgens"***.

II, **Intermediate lobe of Pituitary** : It secretes a hormone called ***intermedin*** or ***"M.S.H."*** [melanocyte stimulating hormone"] It induces the dark coloration in the skin of frog.

III. **Posterior lobe or neurohypophysis of pituitary**: The posterior lobe consists of the axons of neurons of hypothalamus. The two hormones oxytocin and A.D.H. (vasopressin) re formed in hypothalamus, which are stored in posterior lobe. Whenever the body requires, the oxytocin and A.D.H. are released into the blood from the posterior lobe.

(1) **"Oxytocin"** stimulates the contraction of uterus in pregnant woman during the child birth [parturition]. But it is absent in frog.

(2) **A.D.H. [Anti diuretic hormone]** increases the reabsorption of H_2O in the distal convoluted tubule of kidney. Deficiency of A.D.H in creases the urine flow and causes the disease called ***"diabetes insipidus"*** [Diabete -urine, insipid - tasteless]. Hence the urine is highly diluted. Over secretion of A.D.H. increases the absorption of H_2O. Hence the urine is highly concentrated.

THYMUS GLANDS

These are a pair of small, oval reddish bodies lying one on each side just below the tympanum. As the age advances, the glands reduce in size. It secretes a hormone called ***"Thymosin"***. It differentiates the 'T' lymphocytes.

THYROID GLANDS

These are a pair of glands situated very nearer to the external jugular veins in frog. They secrete a hormone called ***"Thyroxine"***. It is rich ***"iodinated hormone"***. It contains and aminoacid called ***"Thyroxine"***. Thyroxine increase the metabolic activities. Hence it maintains B.M.R. [Basal Metabolic Rate]. The thyroxine is essential for metamorphosis in frog. It is also essential for tissue differentiation.

If the thyroid glands ar removed in tadpole larva of frog it fails to undergo metamorphosis. When such thyroid free tadpole larva is fed with thyroid glands, it undergoes metamorphosis. If normal young tad poles are fed with thyroxine, they metamorphose into frogs prematurely [the normal period of metamorphosis is reduced].

PARATHYROID GLANDS

These are four small glands situated very close to the thyroid glands. These secrete calcium hormone called ***"parathyroid hormone"***. It regulates the calcium level in the blood.

ADRENAL GLANDS

There are two, adrenal or supra renal glands situated longitudinally on the ventral surface of kidneys. These are orange colored glands. The outer part of each gland is called ***"cortex"*** and the inner part is called ***"medulla"***. The cortex of adrenal gland is stimulated by A.C.T.H. of pituitary. But medulla is not stimulated by pituitary gland. The cortex secretes ***gluco corticoids, mineral corticoids*** and ***sex corticoids.***

Glucocorticoids : These increase blood glucose level by the synthesis of glucose from proteins. It is called ***"gluconeogenesis"***. Blood glucose level is also increased by the convention of glycogen to glucose. It is called ***"glycogenolysis"***.

Mineral corticoids: These maintain water and salt balance.

Sex corticoids: These stimulate the development of secondary sexual characters.

The central part of medulla of adrenal gland secretes two amines, ***"epinephrine"*** [adrenaline] and ***"norepinephrine"*** [nor adrenaline]. These regulate the blood pressure, heart beat, respirations, contractions and relaxations of smooth muscles. The epinephrine prepares the individual for meeting stress or emergency. Hence the adrenal medulla is considered as the gland for ***"fight, fright and flight"***.

PANCREAS

It is a creamy white colored gland lies between the stomach the duodenum. The gland consists of both ***"exocrine"*** and ***"endocrine"*** parts. Hence the pancreas is described as ***"exo and endo crine gland"***. The exocrine part secretes the digestive juice called pancreatic juice. The endocrine part consists of group of prism shaped cells called ***"islets of Langerhans"***, which lie between the exocrine lobules. The islets of langer hans are of two types, outer *a* [*alpha*] - cells and inner *b* [*beta*] - cells. These *a* and *b* cells produce two types of hormones".

(i) **α-cells** secrete a hormone called ***"glucagon"***. It increases the sugar, glucose level in the blood by glycogenolysis [convention of glycogen into glucose]. Hence glucagon is described as ***"hyper glycemic hormone"***.

(ii) **β-cells** secrete a hormone called ***"insulin"***. [covertion of glucose into glycogen]. So, it lowers the sugar level in the blood. It is described as ***"hypoglycemic"*** hormone.

The glucagon and insulin have opposite actions with one another.

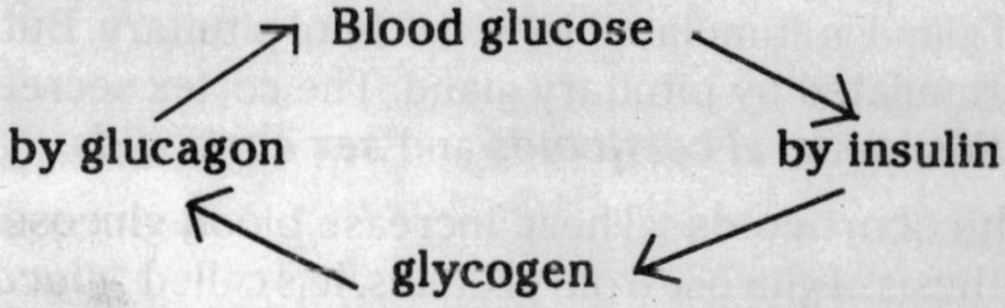

In man the deficiency of insulin causes a disease called ***"diabetes mellitus"*** [sugar disease]. Such patient consists of excess sugar in the blood and eliminates sugar through urine [***"glycosuria"***].

GONADS

The gonads secrete hormones called ***"Sex hormones"***. These are two types, male and female hormones. The gonads are stimulated by pituitary gonadotropins to secrete the hormones.

In male the endocrinal cells of testes are called ***"Leydig cells"***. These are the interstitial cells present between the semi-niferous tubules. The leydig cells secrete male hormones called ***"testosterones"*** or ***"androgens"***. These are responsible for the development of secondary sexual characters.

In female, the follicle cells around the ovum act as the endocrinal cells. These cells secrete female sex hormones called ***"estrogens"***. These are mainly responsible for ***"ovulation"***.

GASTRO-INTESTINAL HORMONES

The mucosa layer of stomach and duodenum acts as endocrinal part. It is made up of columnar epithelium. The mucosa secretes many hormones, which play an important role in the release of digestive juices. The hormones are—

(1) Gastrin: It is secreted by the mucosa of stomach. The gastrin stimulates the gastric glands to secrete gastric juice.

(2) Entero gastrone : It is secreted by the mucosa of duodenum. When the chyme enters the duodenum, the entero gastrone is released into the blood. It reaches the stomach through the blood. Then it inhibits the secretion of gastric juice.

(3) Secretin and Pancreozymin: These are also secreted by duodenum. These stimulate the pancreas to secrete digestive juice.

(4) Cholecystokinin : It is also secreted by duodenum. It stimulates the contraction of gall bladder. Then only the bile juice is released into the duodenum through bile duct.

(5) Entero crinin: It is also secreted by duodenum. It activates the intestinal glands to secrete intestinal juice.

The hormones of alimentary canal regulate the activities glands and maintain the ***"gastro intestinal co-ordination"***.

ENDOCRINOLOGY

Gland	Position	Hormone	Function
1. Pineal body (Epiphysis	on diencephalon	melatonin	protective coloration to tadpole larva
2. Pituitary (anterior)	below the hypothalamus	1. A.C.T.H.	stimulates cortex of adrenal gland
		2. T.S.H.	stimulates Thyroid gland
		3. S.T.H. (Growth) harmone)	controls growth
		4. F.S.H.	regulates
		5. L.H. (I.C.S.H.)	growth of
		6. L.T.H. (prolactin)	gonads and gameto-genesis
Pituitary middle lobe		1. Intermedin (M.S.H.)	Dark coloration in the skin of frog
Pituitary posterior lobe		1. Vasopressin (A.D.H.)	reabsorption of H_2O in kidney
		2. Oxytocin	Contraction of uterus
3. Thymus	below the tympanum	Thymosin	Differentiation of 'T' Lympho cytes
4. Thyroid	nearer to external jugular vein	Thyroxine (metamorph-osing hormone)	regulates general metabolism

(Contd....)

ENDOCRINOLOGY (Contd...)

Gland	Position	Hormone	Function
5. Parathyroid	nearer to thyroid glands	parat hormone	regulates calcium level inblood
6. Adrenal (cortex)	on the ventral surface of kidney	1. glucocorticoids 2. mineral corticoids 3. sex corticoids	regulates gluconeogenesis, minerals in blood
Adrenal (medulla)	———	Adrenaline, nor adrenaline (F.F.F. Hormone)	prepares the individual for meeting stress or emergency
7. Pancreas	between the stomach and duodenum	1. glucagon 2. insulin	glycogenolysis glycogenesis
8. Testis–Leydig cells	anterior to kidney	Testosterones	male secondary sexual characters
ovary (follicle cells)	anterior	estrogens	ovulation
9. Stomach mucosa	inner layer of stomach	gastrin	stimulates the secretion of gastric juice
Duodenum	inner layer	1. entero gastrone	inhibition of gastric juice
mucosa	of duodenum	2. secretin	activates pancreas
		3. cholelcy–stokinin	activates contraction of gall bladder
		4. Entero crinin	activates intestinal glands

11

Reproductive System

Reproductive system is also known as ***genital system.*** The reproductive system consists of principle reproductive organs called ***gonads***. The male gonads are ***testes*** are female gonads are ***ovaries***. Testes produce sperms and ovaries produce ova. The sperms and ova are termed as ***gametes*** or ***sex cells*** in frog, the gametes are released into the water. Hence, fertilization in frog is external.

Sexes are separate in frog and it exhibits sexual dimorphism. The male frog can be easily distinguished from the female frog due to the presence of a pair of vocal sacs and a pair of amplexusory pads. These are well developed during breeding season.

MALE REPRODUCTIVE SYSTEM

The male reproductive system consists of a pair of testes and a pair of urinogenital ducts and cloaca.

TESTES

The male frog consists of a pair of elongated ovoid, yellowish testes situated one on either side of the antero ventral sides of kidneys in the body cavity. Each testis is attached to the antero ventral end of the kidney by a fold of peritoneum called ***mesorchium***. Several delicate tubules called ***vasa efferentia*** arise from the inner margin of each testes. These vasa efferentia pass through the mesorchium and join the ***Bidder's canal*** near the inner edge of the kidney.

The Bidder's canal of the kidney in turn opens into ureter through collecting tubules. Thus the sperms formed in testes enter the kidney through the vasa efferentia, from where they are passed out through ureter.

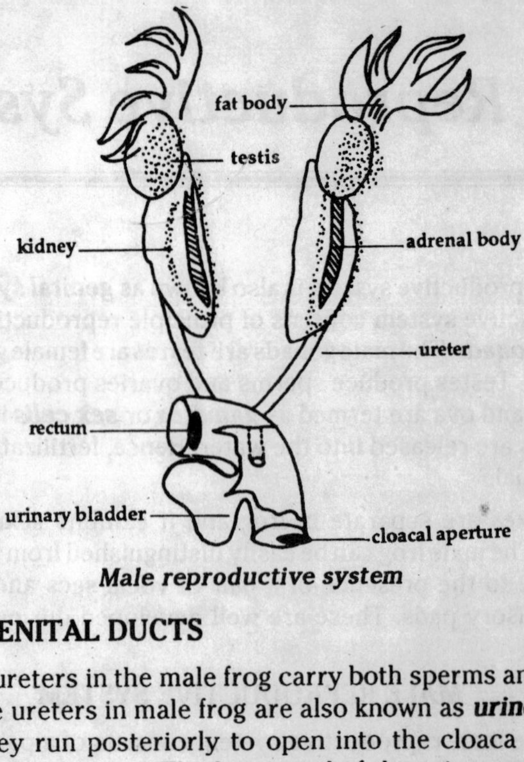

Male reproductive system

URINO-GENITAL DUCTS

The ureters in the male frog carry both sperms and urine. Hence, the ureters in male frog are also known as ***urinogenital ducts***. They run posteriorly to open into the cloaca through urinogenital apertures. The lower end of the urinogenital duct before reaching the cloaca is slightly swollen to form into a ***seminal vesicle***. The sperms are temporarily stored in seminal vesicles.

Fat bodies: Infront of the testes there are several finger-like yellowish coloured processes called ***fat bodies*** or ***corpora adiposa***. The fat bodies serve as reserve food to be used during hibernation and aestivation.

Internal structure of Testis : Each testis consists of a number of coiled tubules called ***seminiferous tubules***. These are held together by a connective tissue.

The connective tissue of testes contains many groups of interstitial cells called ***Leydig cells***. These cells are stimulated by gonadotrophic hormone of pituitary and secrete hormone ***testosterone***. It brings the development of secondary sexual characters like ***amplexusory pads and vocal sacs*** and also sexual behaviour.

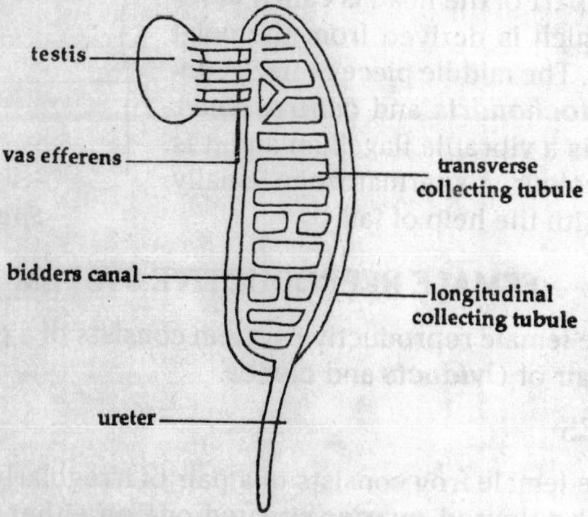

Association of testis with kidney

Spermatogenesis : The inner wall of each seminiferous tubule is lined by a ***layer of germinal epithelial cells.*** These cells produce ***spermatogonia*** (sperm mother cells) which produce sperms. The process of formation of sperms is called ***spermatogenesis***. Bundles of spermatozoa are found in the seminiferous tubules. Each bundle consists of about 25-40 sperms. The sperms are stored temporarily in seminal vesicle. They are discharged into the water through the cloacal opening at the time of amplexus. The mass of sperms discharged by the male frog into water is called ***milt***.

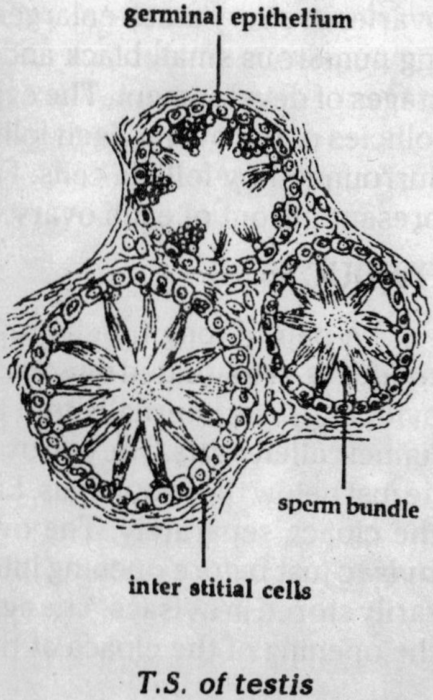

T.S. of testis

Sperms or Spermatozoa : Each mature sperm measures about 0.03 mm in length. It consists of three parts, namely 1. head, 2. neck or middle piece and 3. tail. The head of the sperm contains a large nucleus and ***acrosome***. The anterior part of the head is called ***acrosome*** which is derived from the golgi complex. The middle piece or neck contains ***mitochondria*** and ***centrosomes***. The tail is a vibratile flagellum and it is grey in colour. A spermatozoan usually moves with the help of tail.

Sperm

FEMALE REPRODUCTIVE SYSTEM

The female reproductive system consists of a pair of ***Ovaries***, a pair of ***Oviducts*** and ***cloaca***.

OVARIES

The female frog consists of a pair of irregularly lobed sac like, dark coloured ***ovaries*** situated one on either side in the body cavity on the ventral surface of the kidneys. Each ovary is made up of a number of lobes and is attached to the antero ventral side of kidney by a fold of peritoneum called ***mesovarium***. The ovaries are very much enlarged during breeding season containing numerous small black and white ova which are in different stages of development. The eggs or ova are developed in minute follicles of the ovary. Each follicle consists of an enlarged ovum surrounded by follicle cells. Fat bodies or corpora adiposa are present infront of each ovary.

OVIDUCTS

A pair of long tubular and highly convoluted tubes called ***oviducts***, are situated one on either side in the body cavity. Each oviduct opens internally into the body cavity by a wide ciliated funnel called ***ostia***. The two oviducal funnels face each other and lie just below the esophagus. Each oviduct opens posteriorly into the cloaca separately. The oviduct is enlarged to form a wide ***ovisac*** just before opening into the cloaca. The eggs are temporarily stored in ovisacs. The eggs are released into water through the opening of the cloaca at the time of copulation.

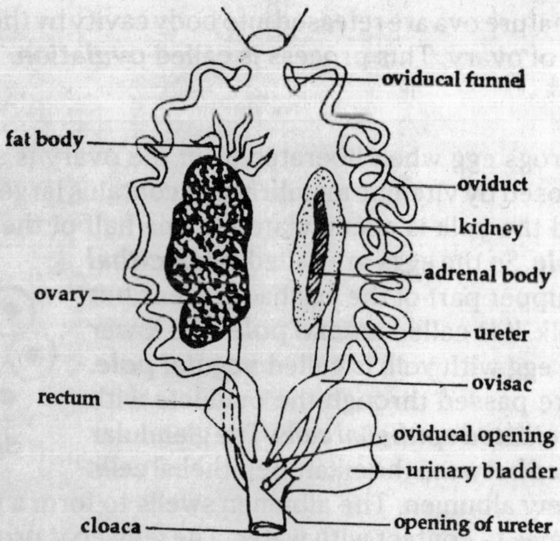

Female reproductive system

OOGENESIS

The wall of the ovary is made up of an outer layer of ***connective tissue and an inner layer of germinal epithelial cells.*** The process of formations of eggs is called ***oogenesis.*** The follicles of the ovary are formed from germinal epithelial cells. The eggs are actually produced by the repeated divisions of the germinal epithelial cells.

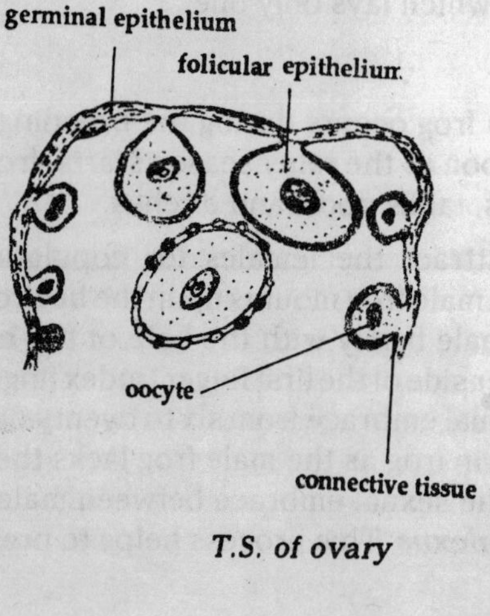

T.S. of ovary

The mature ova are released into body cavity by the rupture of the wall of ovary. This process is called ***ovulation***.

OVA

The frogs egg when liberated from the ovary is spherical and is enclosed by vitelline membrane. It contains large quantity of yolk and the yolk is accumulated in one half of the egg i.e.., ***vegetal pole.*** So the eggs are called ***telolecithal type***. The upper part of the egg has nucleus but without yolk. It is called ***animal pole.*** The lower part of the egg with yolk is called ***vegetal pole.*** The ova are passed through the oviducts with the help of *ciliated epithelial cells.* The glandular cells situated between the ciliated epithelial cells secrete slimy albumen. The albumen swells to form a jelly coat when it comes in contact with water. The jelly coat protects the eggs from injuries. It keeps the eggs warm by allowing the sun rays. The mass of eggs released by female frog into the water is called ***frog's spawn.***

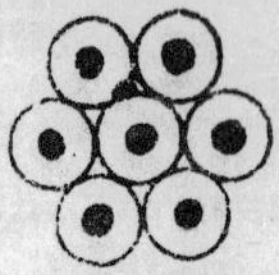

Spawn

Larger frogs, as a rule produce greater numbers of eggs. A bullfrog may lay as many as 20,000 eggs. Smaller species, such as the green frog and leopard frog, lay up to 6,000. Toads produce smaller eggs than frogs: these eggs are much more numerous. A toad the size of leopard frog may lay more than 25,000 eggs. The eggs of tree frogs are usually numbered in the hundreds. The record for the smallest number of eggs belongs to a West Indian form Sminthillus, which lays only one.

3. COPULATION

Copulation in frog occurs during the breeding season i.e., rainy season. As soon as the rainy season starts, frogs return to the water of ponds, tanks pools and ditches.

Male frogs attract the females for copulation by their ***croaking sound***. A male frog mounts upon the back of the female and clasps the female firmly with the help of the ***nuptial pads*** present on the inner side of the first finger (index finger). The pair remains in this sexual embrace from six to twenty days. There is no real copulation in frog as the male frog lacks the copulatory organ. There for the sexual embrace between male and female frogs is called ***amplexus***. This process helps to press the abdomen of the female by which the eggs are discharged into water through the cloacal aperture. The laying of eggs by female during this copulatory act is called ***spawning***. As soon as the eggs are discharged by the female into water, the male sheds its sperms over them.

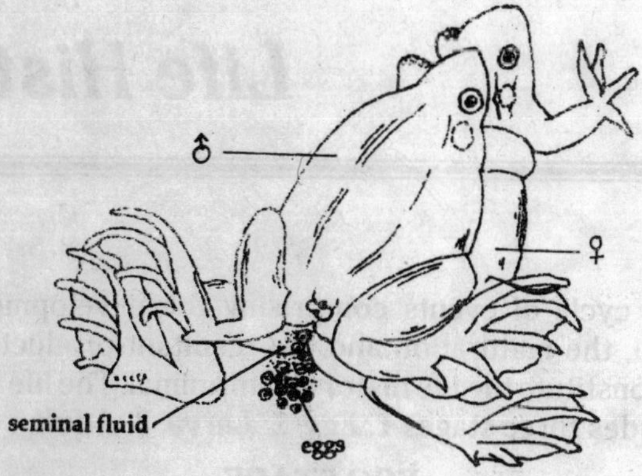

Amplexus

Fertilization: The fusion of the male and female gametes is called ***fertilization*** and the fertilized eggs are called ***zygotes***. In case of frog, the fertilization takes place in water. Hence, the type of fertilization is called ***external fertilization***.

12

Life History

The cycle of events comprising the development of an individual, the maturation and the resultant production of off springs constitute the life history of an animal. The life history of frog includes three stages 1. ***Egg***, 2. ***Larva***, 3. ***Adult***.

EGG STAGE

Development starts soon after the fertilization. The zygote undergoes repeated divisions to become an embryo. The embryo develops into a fish-like ***Tadpole larva*** which undergoes metamorphosis to become into an adult. It is not a simple process but involves a series of stages resulting in a gradual increase in the structural complexity. The study of development of animals from fertilized egg is called ***embryology***. The development of a multicellular organism starts with a single celled structure called ***zygote***.

ZYGOTE

It is about 1.6 mm. in diameter. The upper half is known as ***animal pole*** which is black and pigmented. It contains the nucleus and cytoplasm. The lower half is known as ***vegetal pole*** which is unpigmented, pale or light yellow in color and contains full of yolk. In between these two regions there is a small area known as ***greycrescent*** which is situated in the equatorial region opposite to the sperm entrance point. The greycrescent indicates the future dorsal side of the embryo.

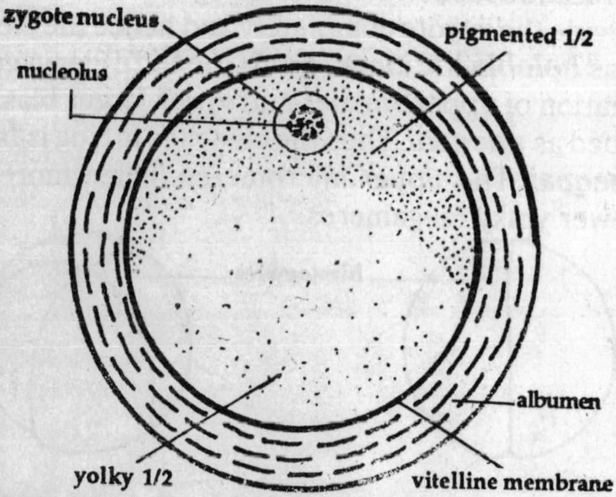

Zygote

As the yolk is present in moderate quantity, the egg of frog is described as ***mesolecithal*** and as the yolk is distributed towards one side of the egg it is described as ***telolecithal***. Hence the egg of frog is described as ***mesolecithal*** and ***telolecithal***.

The division of the zygote is called ***cleavage*** or ***segmentation***.. By cleavage single celled zygote divides into multicellular embryo. The resulting cells of the cleavage are called ***blastomeres***. The zygote undergoes cleavage in few hours after fertilization. The first cleavage is ***vertical*** or ***meridional***. The cleavage furrow starts from the animalpole and reaches to the vegetal pole thus dividing the egg completely into two blastomeres. The second cleavage is also ***vertical*** or ***meridional*** but at right angles to the first cleavage resulting in the formation of four blastomeres. The third cleavage is ***latitudinal*** or ***horizontal*** occurring above the equator resulting in the formation of 8 blastomeres. The 8 blastomeres are arranged in two tiers of four each. As the cleavage furrow is nearer to the animalpole the resulting upper four blastomeres are smaller than the lower four blastomeres. The upper smaller blastomeres are described as ***micromeres*** and are situated in the animalpole. The lower larger blastomeres are called ***macromeres*** or ***megameres*** and are situated in vegetalpole. The megameres are filled with yolk so they are also called ***yolk cells***.

The first and second vertical cleavages and third horizontal cleavage divide the zygote completely and hence the cleavage is described as ***holoblastic cleavage***. As the third cleavage results in the formation of upper smaller and lower larger blastomeres it is described as ***unequal***. Thus the cleavage in frog is ***holoblastic*** and ***unequal***. The upper micromeres divide more rapidly than the lower yolky megameres.

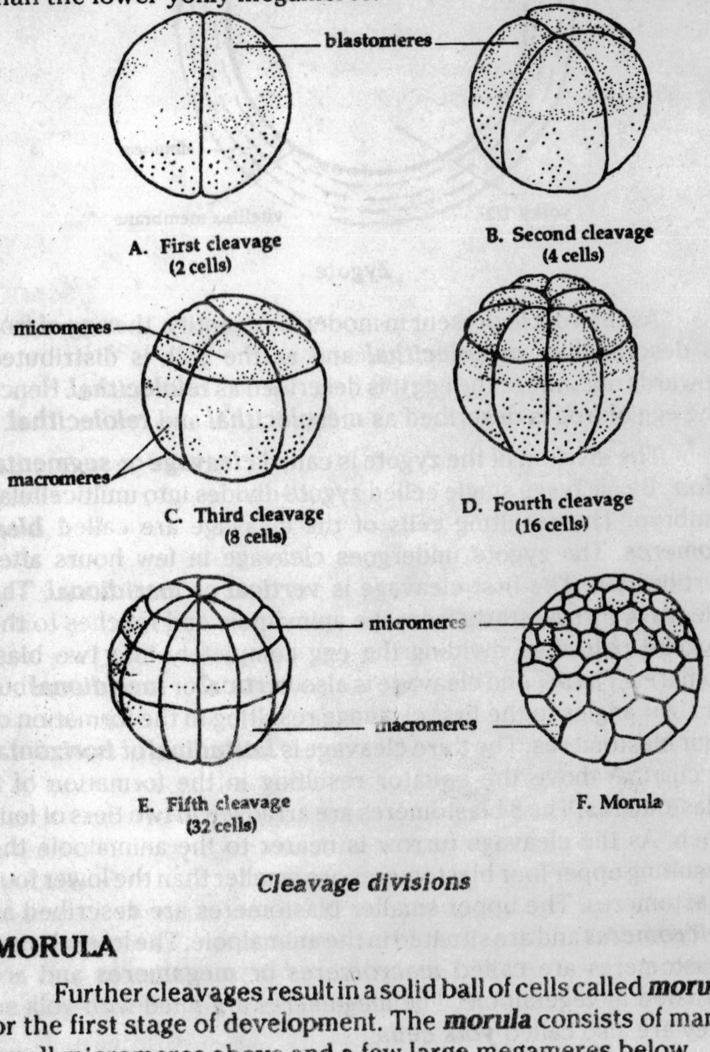

Cleavage divisions

MORULA

Further cleavages result in a solid ball of cells called ***morula*** or the first stage of development. The ***morula*** consists of many small micromeres above and a few large megameres below.

BLASTULA

It is the second stage in the development of an organism. It is formed from the morula. The process of formation of blastula is known as ***blastulation***. By repeated cleavages the number of blastomeres increases and a cavity appears inside the embryo. This is known as ***segmentation cavity*** or ***blastocoel***. This cavity is lined by a single layer of blastomeres. Thus a single layered hollow ball like structure enclosing a cavity known as blastocoel, is formed during development and this is known as ***blastula stage***. The blastocoel in frog is slightly excentric as it is situated towards the animal pole. The roof of the blastocoel is made up of several layers of micromeres while the floor is composed of yolky megameres.

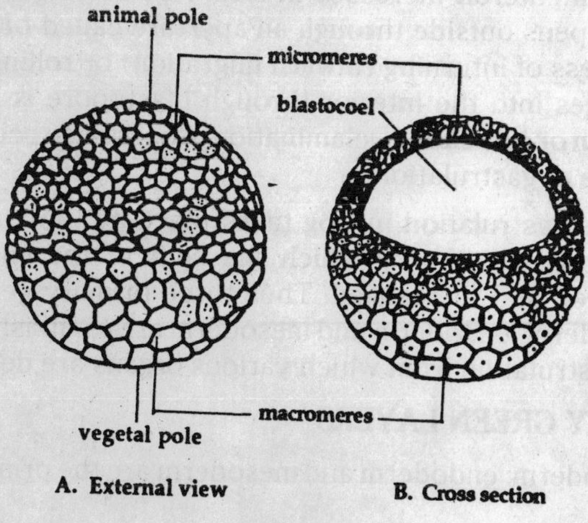

Blastula

GASTRULA

Gastrula is the third stage in the development of an embryo. The process of formation of the gastrula from blastula is known as ***gastrulation***. Gastrula is a double walled structure consisting of an outer ***epiblast*** (ectoderm) and an inner ***hypoblast*** (endoderm) enclosing a cavity known as ***primitive gut*** or ***arch-enteron*** or ***gastrocoel***. During gastrulation, rearrangement and reorganization of cells take place. Gastrulation is also marked by a series of mass migration of cells which could be termed as ***formative movements***.

In frog gastrulation takes place by ***epiboly, invagination, involution*** or ***intucking*** and ***delamination***. The micromeres of animal pole which have no yolk divide more rapidly and actively than the yolk ladden megameres of vegetal pole. As a result, the micromeres grow over the megameres covering them completely except a small area called ***yolk plug***. This process of overgrowth of micromeres over megameres is known as ***epiboly***. At this stage, a small crescentic groove appears on the dorsal side near grey crescent by invagination. This groove is the beginning of ***archenteron***. It grows and deepens into the interior and forms into a cavity called ***archenteron*** or ***primitive gut***. As the archenteron grows into a large cavity, the blastocoel gets reduced. The megameres are pushed towards the blastocoel, by which archenteron increases in size. At this stage, the archenteron opens outside through an aperture called ***blastopore***. The process of inturning (inward migration) or rolling of outer micromeres into the interior through blastopore is known as ***involution or intucking***. Delamination process also occurs in the final stage of gastrulation.

Thus gastrulation in frog takes place by the above mentioned processes during which the cells of the embryo are reorganized and rearranged. Thus three primary germ layers *viz.*, ectoderm, endoderm and mesoderm are established by the end of gastrulation from which various organs are derived.

PRIMARY GREEN LAYERS

Ectoderm, endoderm and mesoderm are the primary germ layers.

Ectoderm

This is the first germ layer. It gives wing organs:

> Epidermis and derivatives like cutaneous glands, nervous system (brain, spinal cord and nerves), Eye (retina, lens, cornea, conjunctiva), olfactory organs (nasal epithelium) and membranous labyrinth (internal ear), lining of stomodaeum (buccal cavity) and proctodaeum (rectum), Chromatophores (pigment cells), and lateral line sense organs, Pineal gland (epiphysis), posterior and intermediate lobes of pituitary gland and medullary portion of the adrenal gland.

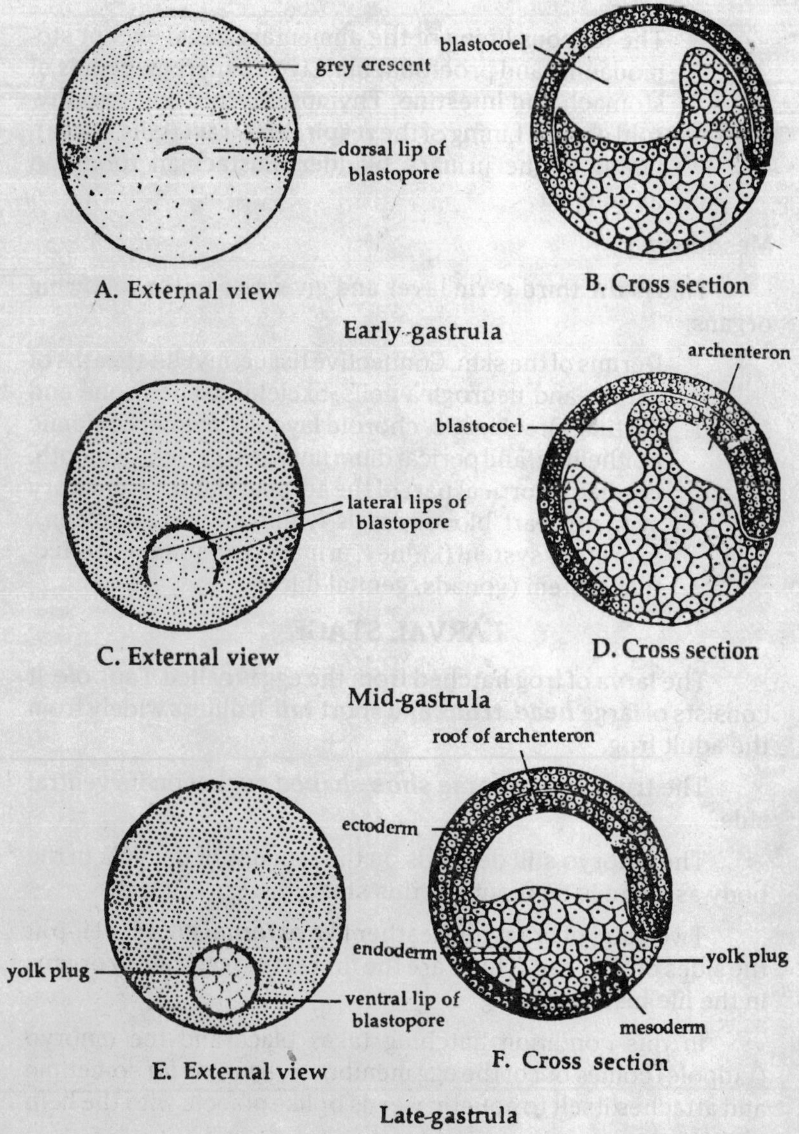

Gastrulation

Endoderm

This is the second germ layer and gives rise to the following organs:

The mucous lining of the alimentary canal except stomodaeum and proctodaeum, Liver, pancreas, glands of stomach and intestine, Thymus, thyroid and parathyroid glands, Lining of the respiratory tract (gills, lungs), Lining of the urinary bladder, Eustachian tube and middle ear.

Mesoderm

This is the third germ layer and gives rise to the following organs:

Dermis of the skin, Connective tissue, myelin sheaths of axons, and neuroglea cells, Skeletal tissue (bone and cartilage). sclerotic, choroid layers, Coelom, coelomic epithelium, and pericardium muscles (skeletal, smooth, cardiac), Cortical part of the adrenal gland, Circulatory system (heart, blood vessels, lymphatic system spleen), excretory system (kidney, urinary ducts and reproductive system (gonads, genital ducts).

LARVAL STAGE

The larva of frog hatched from the egg is called ***Tadpole***. It consists of large ***head***, ***trunk*** and short ***tail***. It differs widely from the adult frog.

The trunk bears ***horse shoe shaped sucker*** on its ventral side.

The embryo still depends on the remains of the yolk in the body as there is no mouth at this stage.

Two pairs of branched feather like ***external gills*** develop at the sides of the head. These are the first set of respiratory organs in the life history of frog.

In this condition hatching takes place and the embryo (tadpole) comes out of the egg membrane. It swims for sometime and attaches itself to aquatic weeds or like objects with the help of its sucker.

The trunk at its anterior end bears a pair of prominences and a pair of depressions which later develop into eyes and nostrils of the adult frog.

The larva has a gut in which yolk is still present.

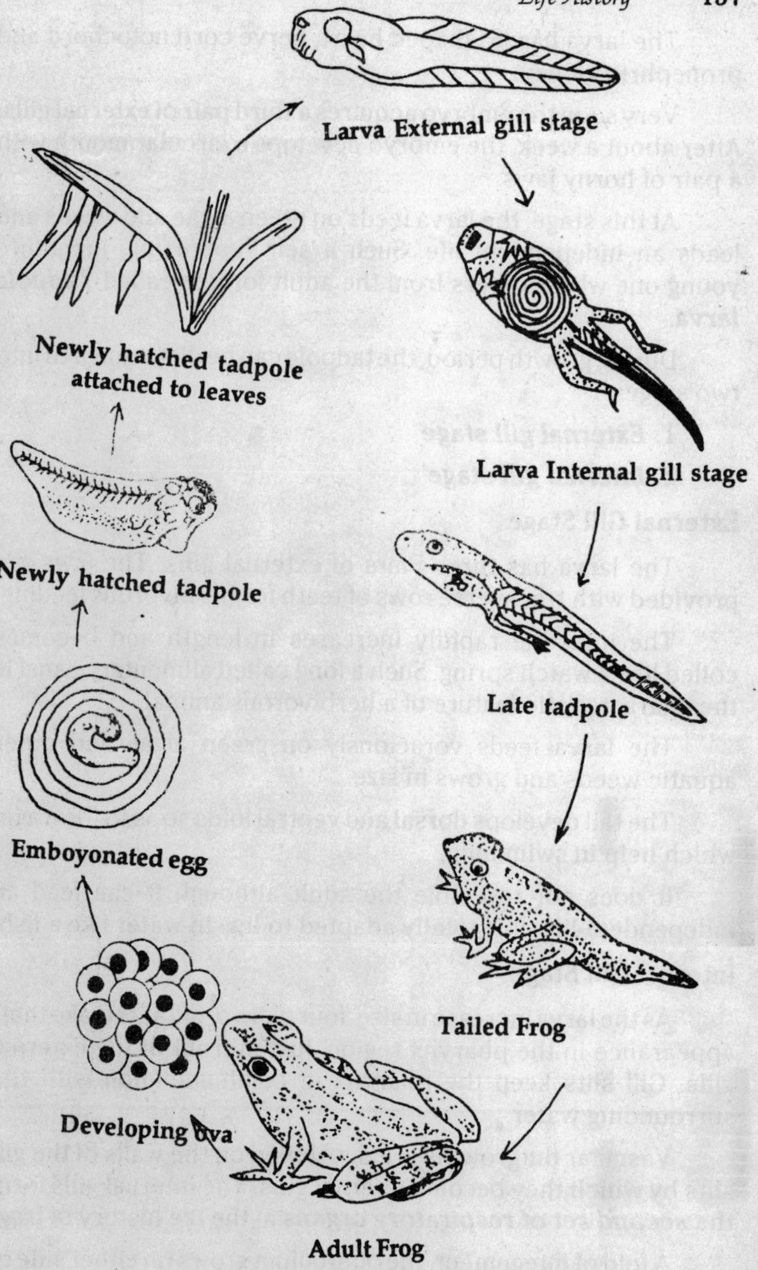
Larva External gill stage
Larva Internal gill stage
Late tadpole
Tailed Frog
Adult Frog
Developing ova
Emboyonated egg
Newly hatched tadpole
Newly hatched tadpole attached to leaves

Life Cycle

The larva has 'S' shaped heart, nerve cord notochord and pronephric kidney.

Very soon the embryo acquires a third pair of external gills. After about a week, the embryo develops a circular mouth with a pair of horny jaws.

At this stage, the larva feeds on green algae and weeds and leads an independent life. Such a self supporting, immature young one which differs from the adult form is called ***Tadpole larva.***

During growth period, the tadpole can be distinguished into two stages.

1. ***External gill stage***
2. ***Internal gill stage***

External Gill Stage

The larva has three Pairs of external gills. The jaws are provided with transverse rows of teeth for herbivorous feeding.

The intestine rapidly increases in length and becomes coiled like a watch spring. Such a long called alimentary canal is the characteristic feature of a herbivorous animal.

The larva feeds voraciously on green algae and other aquatic weeds and grows in size.

The tail develops dorsal and ventral folds so as to form a fin which help in swimming.

It does not resemble the adult although it can lead an independent life. It is totally adapted to live in water like a fish.

Internal Gill Stage

As the larva increase in size, four pairs of gill slits make their appearance in the pharynx region, just in front of the external gills. Gill slits keep the pharynx in communication with the surrounding water.

Vascular outgrowths are developed on the walls of the gill slits by which they become internal gills. The internal gills form the ***second set of respiratory organs*** in the life history of frog.

A fold of integument, the operculum grows on either side of the head infront of the gill clefts. It extends backwards and covers and four gills of its side. The two opercular folds meet ventrally and their posterior margins fuse with the body wall leaving an opening called the ***spiracle***. The chamber enclosing the gills is called ***branchial chamber***. The water entering the opercular cavity passes out through this opening.

Both the external and internal gills function for sometime and later the external gills begin to disintegrate and finally disappears with the growth of opercular folds.

The respiratory gaseous exchange occurs when the water passes over the gills. Tadpole at this stage respires exactly like a fish.

By this time the lateral sense organs are formed on either side of the body and the larva resembles fish.

The circulatory system of tadpole is also like that of a fish. The sucker is lost and eyes are formed. This stage of tadpole is called ***internal gill stage***.

Other Changes in the Larva

Mean while, the fore limbs and the hind limbs appear as small buds and gradually increase in size. The fore limbs are hidden under the operculum and not visible outside till they are fully grown.

Hind limbs arise first as buds at the base of the tail. Both pairs of limbs lengthen by developing all the structures, but the hind limbs are very stronger and longer.

Later on one pair of lungs arise as ventral outgrowths of the pharynx. They communicate outside through the nasal passages and external nostrils.

The larva respires by both lungs and gills for sometime later the internal gills begin to disappear and lungs become the main respiratory organs.

The circulatory system undergoes changes with the appearance of lungs, i.e., to suit the pulmonary respiration.

The horny jaws are lost. The upper and lower jaws are formed. Mouth becomes wide. The tongue grows in size.

The alimentary canal becomes shorter and straight with few coils. Eyes become prominent. Liver and pancreas get enlarged.

The animal changes its food habit, i.e., it becomes carnivorous, feeding on insects, worms, etc.

Fore limbs become more conspicuous as the operculum fuses with the skin. The ciliated larval skin is shed off as the new skin is formed.

There is an increase in the production of thyroxine hormone towards the end of the tadpole phase. With the increasing differentiation of limbs the tadpole frequently comes out of water and leaps on the land.

The pronephric kidney disappears and mesonephric kidney is formed. The nitrogenous excretory waste changes from ammonia to urea.

The body becomes differentiated into a triangular head and a broad trunk with two pairs of limbs. After undergoing all these changes, the young frog leaves the water for dwelling in damp places and in course of time it becomes adult.

The tail gradually gets reduced and absorbed. The lysosomes having hydrolyzing, enzymes are responsible for the absorption of the tail.

The larva feeds up on its tail by phagocytic action.

METAMORPHOSIS

A fish-like, aquatic gill breathing, herbivorous, tailed larva is transformed into an amphibious lung breathing, carnivorous, tailless frog. This phenomenon is called metamorphosis. Tadpole does not feed at the time of metamorphosis. The series of changes that take place during the transformation of larva into adult is called metamorphosis. Thyroxine, a hormone from thyroid gland initiates the metamorphosis in Tadpole.

The presence of a fish-like tadpole stage in the life history of frog indicates the evolution of amphibians from fish-like ancestors. The tadpole is believed to be recapitulating the ancestral features.

The length of time between egg-lying and metamorphosis varies greatly among different species of frog. Some such as the spadefoot toads, which lay their eggs in temporary pools in desert regions, may go through the cycle in less than two weeks. At the other extreme in cold northern climates the bullfrog may

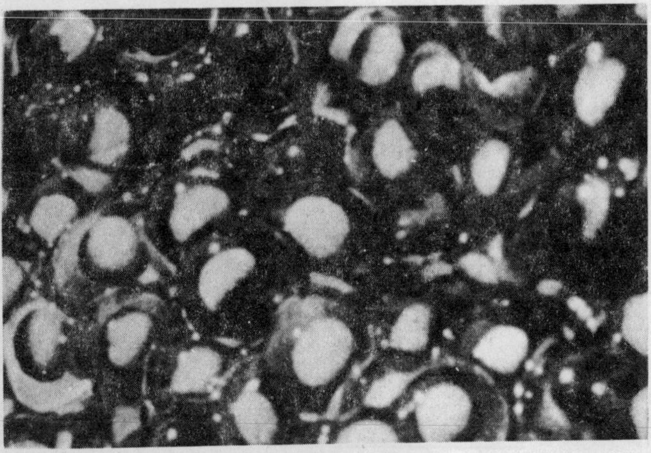

Frog Eggs

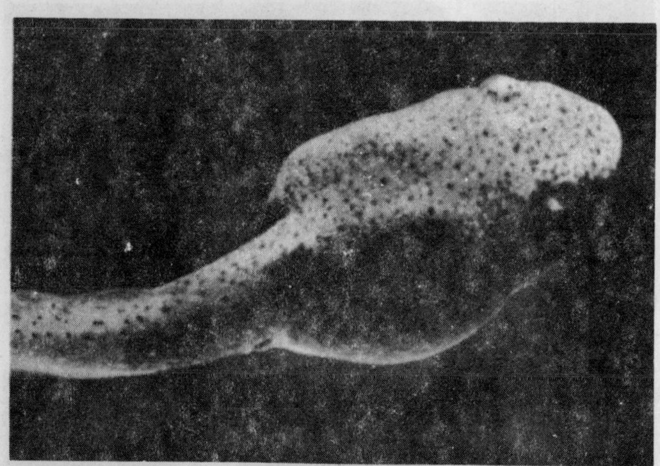

Tadpole larva

Tailed Frog

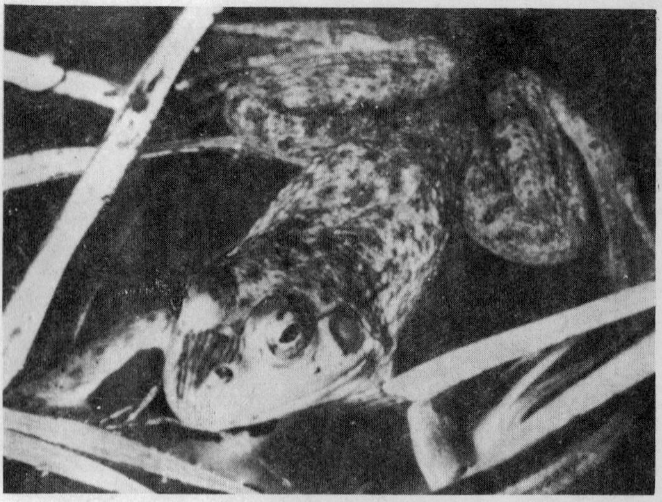

Adult Frog

LIFE HISTORY OF FROG

take two or three years to transform. For the majority of species the transformation is completed during a single season. The eggs laid in spring or early summer produce small frogs by the end of summer or early fall.

UNUSUAL DEVELOPMENT

The development of certain frogs does not follow the general pattern just described. In various species, the tadpole is not free living form. It remains in the eggs, and metamorphosis takes place before the egg hatches. Such eggs are never laid in the water, but only in moist places on land or in trees.

The male of one remarkable South American species, *Rhinoderma darwini*, keeps the eggs in his vocal pouch. The female Surinam toad. *Pipa pipa*, carries her eggs about on her back. The eggs are set in place by the male. They sink into the skin and each is protected by a covering. The larva that develops from each egg completes its development in its tiny chamber and emerges as a small but fully formed toad.

The female midwife toad. *Alytes obstetricans*, lays two strings of eggs. Her mate thrusts his legs through the egg mass. Thereafter, he carries the eggs, attached to his legs, about with him until the larvae hatch out. By that time the larvae are well developed.

WE CAN OBSERVE THE LIFE HISTORY OF FROG IN OUR LIVING ROOM

We can see for ourselves how a tadpole develops into a frog by collecting some frogspawn from a pond. The best time to collect is when the weather is getting warmer at the start of spring. At this time the frogs wake up from their long sleep during the cold winter months and go to the ponds to breed. It is only necessary to take a small amount of the frogspawn, just enough to cover the bottom of a small jar. If we look after it, lots of frogs will hatch out.

Put some pond water in the jar too. When you get home put the jar in a cool place. If the temperature is too hot the eggs will die. Tiny black tadpoles will hatch out of the eggs in a few weeks. The time varies according to the species of frog whose eggs we are studying.

Put some of the tadpoles into other jars to avoid crowding. Feed them bits of leafy vegetation. If the pond is local, some of the pond weed there would be the best food. They will also eat the yolk of a hard-boiled egg. If we observe them closely, we will be able to see changes in the tadpoles. At first they have gills, and long tails. Then the hindlegs emerge followed soon after by the forelegs. The gills will have disappeared by this stage. Soon the tail gets shorter and eventually disappears. The tadpole has changed into a frog. We will have to be patient to see these changes.

It may take a year (or two in the case of the bullfrog) for the frog to emerge.

We can also see the change from water to land life if we build a semi-aquatic tank. Put a shallow bowl of water at one end of the tank. At the other end make a landscape of rocks, plants, sand and soil. When the tadpoles have well developed lungs they will leave the water and jump out onto the land.

When we have observed these things, we have to take the frogs back to a near-by pond or stream so they can live in their natural environment.